AF572631

Virus Separation and Purification Methods

Virus Separation and Purification Methods

edited by

Alfred Polson
Tygerberg Hospital and
University of Stellenbosch
Stellenbosch, South Africa

Marcel Dekker, Inc. **New York • Basel • Hong Kong**

Library of Congress Cataloging-in-Publication Data

Virus separation and purification methods / edited by Alfred Polson.
p. cm.
"Originally published as Preparative biochemistry, volume 23.
numbers 1&2, 1993"–T.p. verso.
Includes bibliographical references and index.
ISBN 0-8247-9149-5 (alk. paper)
1. Viruses–Isolation. I. Polson, Alfred.
[DNLM: 1. Viruses–isolation & purification. QW 160 V821169
1993]
QR385.5.V57 1993
576'.64'028–dc20
DNLM/DLC
for Library of Congress 93-22933
CIP

The contents of this volume were originally published as *Preparative Biochemistry*, Volume 23, Numbers 1&2, 1993.

The publisher offers discounts on this book when ordered in bulk quantities. For more information, write to Special Sales/Professional Marketing at the address below.

This book is printed on acid-free paper.

MARCEL DEKKER, INC.
270 Madison Avenue, New York, New York 10016

Current printing (last digit):
10 9 8 7 6 5 4 3 2 1

PRINTED IN THE UNITED STATES OF AMERICA

Contents

PART II - SEPARATION AND PURIFICATION OF SPECIFIC VIRUSES

Acknowledgments

The editor is grateful to Professor T. F. Kruger, Director of the Infertility and Reproductive Biology Unit, for his unfailing interest in the compilation of the works and to Mrs. Julia Husselmann for the large amount of typing involved.

Preface

In the course of his sixty-year-long (and still ongoing) scientific career, Dr. Polson has contributed more to the large field of characterization and purification of proteins, viruses, antibodies and biopolymers in general than anyone now alive (or dead). While many of his contributions are well-known (e.g., the polyethylene glycol approach to protein fractionation), even though it may not be widely realized that Polson was the inventor, other aspects of his work, especially on virus purification, may be known only to small groups of scientists.

Many of his purification approaches can be of great help to virologists, immunologists and biochemists. We therefore feel that a concentrated collection of Polson's purification methods (many of which were published earlier in *Preparative Biochemistry*) strongly merit re-publication in book form.

Following a biographical sketch, Dr. Polson's papers are organized in two parts: Part I-*Methodology*, and Part II-*Separation and Purification of Specific Viruses*. Within each part, the papers are ordered chronologically.

Carel J. van Oss

Biographical Sketch Of Dr. Alfred Polson

Dr. Polson was born on a farm in the District Brandfort, Orange Free State, South Africa, March 11, 1912. He obtained the B.Sc. degree from the University of Stellenbosch in November 1932. Subjects studied were chemistry, geology, physics and mathematics. In November 1933 he specialized in physical chemistry and obtained the M.Sc. in November 1933, on the subject: "The Influence of Water Vapour on Gas Reactions."

He proceeded to Sweden in January 1934 and did research for a doctorate at the Institute for Physical Chemistry, University of Uppsala. The director of the institute at that time was (Nobel Prize winner) Prof. The Svedberg. The title of his thesis was "Researches on the Diffusion Constants of Proteins"; it was completed in March 1937 and was published in full in *Kolloid Zeitschrift* in 1939. He returned to South Africa in September 1937 and in November of the same year was awarded the D.Sc. degree at the University of Stellenbosch.

In January 1938 Dr. Polson joined the staff of the Institute of Veterinary Research at Onderstepoort, Transvaal, South Africa, where he did research on the viruses peculiar to domestic animals in South Africa and also assisted in the development of vaccines against these viruses. He developed a very effective vaccine against botulism in bovines by growing the infectious organism in cellophane dialysis tubing filled with saline and suspended in the culture medium. This work was done in 1945.

In January 1946 Dr. Polson did research in Prof. J. W. Williams' Department of Chemistry of the University of Wisconsin, Madison, Wisconsin. The subject of his research was "Electrophoresis of the Venom of Rattle Snakes" (*Crotalus* species) and identification of the toxin amongst the components of the venom. From July 1946 to April 1948 he did research at the N.I.H., Bethesda, with Dr. Ralph Wyckoff. The subjects of his research there were quantitative paper chromatography of the amino acids in lysates of *E. coli* organisms and their bacteriophages, on the abnormally high diffusion rates of the bacteriophages of *E. coli* and on the electron microscopy of the haemocyamin of the horseshoe crab. He held an N.I.H. scholarship during the period July 1946 to April 1948.

Dr. Polson returned to Onderstepoort and continued his work on animal viruses until May 1952 when he was appointed Research Officer (Virology) in the Virus Research Unit of the Council for Scientific and Industrial Research (CSIR). His work there concerned the viruses of poliomyelitits; their purification and physico-chemical characterization as well as work on the safety of formaldehyde-inactivated polio vaccine. Other viruses which he studied were several arbor viruses, rabies and influenza virus. During 1956 to 1957 he worked at the Karolinska Institute, Stockholm, with Prof. Sven Gard, continuing his work on poliomyelitis virus.

Dr. Polson returned to the CSIR Virus Research Unit in September 1957 and continued his work on polio and other enteroviruses as well as influenza virus. He developed an effective influenza vaccine by aggregation of the virus with polyethylene glycol, followed by inactivation with formaldehyde. He also developed two rotors for the Beckman preparative ultracentrifuge. These were the thin layer and the virus extraction rotors. In 1958 he obtained a second D.Sc. degree (Cape Town University) on papers published (1939-1958). In 1963 Dr. Polson was invited as guest worker by Prof. R. Porter of the Department of Immunology, St. Mary's Hospital, London, for the period August 1963 to August 1964.

Upon reaching his retiring age in 1975, Dr. Polson accepted a Senior Lecturership in Microbiology at Cape Town University. Apart from his teaching duties he did research on plant viruses. He introduced the technique of immunizing chickens and extracting the antibodies from the yolks of eggs laid by the chickens, using the polyethylene glycol procedure. In January 1980 he accepted a research post in the Department of Biochemistry, University of Stellenbosch, where he continued improving the method of separating antibodies from the yolks of eggs from immunized chickens. He also introduced the method of preparative ultracentrifugation in tubes filled with uniform beads to reduce the time of centrifugation of proteins and viruses from suspension into concentrates or pellets by 90% or more, depending upon the size of the beads used. At the beginning of 1988 Dr. Polson was appointed as member of the research group in the Infertility and Biological Reproductive Unit of the University of Stellenbosch, Tygerberg Hospital, South Africa.

Dr. Polson has published altogether 180 to 200 papers in the following journals: *Nature*, *Science*, *Bioch. J.*, *Experientia*, *Biochim. et Biophys.*

Acta, *Kolloid Zeitschrift*, *Proc. Soc. Exp. Biol. and Med.*, *J. Hyg*, *Prep. Biochem.*, *Immunol. Invest.*, *Onderstepoort J.* and *SA Journal J. Sci.*

HONORARIA

1975: A gold medal from the Medical Research Council of South Africa.
1981: An award from the Claud Harris Leon Foundation.
1987: Ph.D. (Hon.) from the University of the Orange Free State.
1988: A gold medal from the South African Veterinary Medical Society.

Part I - Methodology

1. Purification of Viruses by Electrophoresis in Sucrose Concentration Gradients*

ABSTRACT

Zone electrophoresis (ZE) in concentration gradients of sucrose proved to be a powerful technique in the purification of viruses although it is seldomly used as a single step for the separation of the infective agent from extraneous matter but usually as a final procedure. No difficulties were experienced in the purification of plant viruses with ZE possibly because of the resistance of the agents to the chloroform butanol treatment which the impure plant viruses received prior to the ZE. Likewise no difficulties were encountered during purification of human and animal picorna viruses because they also are resistant to chloroform or chloroform-butanol treatment prior to ZE.

Purification of animal and possibly human reoviridae ZE as a final step is involved as they are associated with or even attached to the host cell membranes which are lipoidal in nature. Because the lipoidal material on the virus has an affinity for organic solvents the virus particles are caught in the emulsion when an extract of the crude virus material is shaken with the organic solvent. With these

*A. Polson. Data in part from A. Polson and B. Russell, in: Methods in Virology (Maramorosch and Koprowski, eds.), and from I. R. M. Juckes, Ph.D. thesis, U. of Capetown, 1972.

viruses polyethylene glycol Mr 6000 (PEG) is often helpful as a means of displacing most of the host proteins as a step in the partial purification prior to ZE. Neurotropic African horse sickness virus which presented many problems to overcome was rendered free of extraneous proteins to the extent that electron micrographs could be made of it. Neurotropic Rift Valley fever virus was similarly difficult to purify as it was always associated with mouse brain components. This virus was finally purified from infected mouse blood in which the virus particles were not associated with host cell debris and other extraneous proteins.

Zone electrophoresis experiments were not conducted on viruses with large particle sizes e.g. myxoviruses because they could be purified without a great deal of effort using polymer displacement, and thin layer centrifugation. Zone electrophoresis as a means of final purification was vividly displayed with the purification of five viruses of the larvae of the pine emperor moth, *Nudaurelia cytheria cytheria*.

INTRODUCTION

The phenomenon of electrophoresis has proved to be valuable to plant virologists in identifying plant viruses. Evidence of the presence of one or more infected agents in a diseased plant may be easily established. This information is especially important since different viruses may evoke similar symptoms in the same plant illustrating the danger of relying only on symptoms in identification of the causative agent of a plant disease.

Animal viruses have also been investigated by electrophoresis but, because of different factors such as the association with many proteins, preliminary purification of the infected agents is essential prior to electrophoresis. Due to their usual unstable nature especially of the viruses of particle diameters >50 nm preliminary purification is usually accompanied by considerable loss of viability. By conducting electrophoresis in the presence of proteins which reduce surface tension, the losses in viability may be considerably reduced.

The difficulties associated with the moving-boundary method led to the development of zone electrophoresis (ZE) in concentration gradients of inert substances. In ZE in concentration gradients hydrodynamic stability is insured by the density gradient which parallels the concentration gradient. An important factor is that the components in the mixture subjected to ZE migrate from a thin layer into the gradient at different rates. The layers widen when the migration proceeds from high density gradient towards low density. When the electrophoresis proceeds from low density gradient towards high density, the zones maintain their original widths or may even be condensed. The reason for this effect is that the front of the zone is retarded relative to its rear because of increase of viscosity of the medium across the migrating zone. The reverse holds for migrations from a region of high to lower concentration.

Svensson and Valmet[1] introduced a zone electrophoresis apparatus that was designed for use with solid media. The apparatus was later converted for ZE in concentration gradients. The apparatus was used for purification of F.A. mouse encephalitis virus[2]. The apparatus was cumbersome to use routinely and a less complicated design was introduced. The new apparatus could be easily assembled, sterilized and is easy to use.

Description and Construction of the Zone Electrophoresis Apparatus

The ZE apparatus[3] that meets most requirements of virologists and protein chemists interested in separation techniques is shown in Fig. 1. The apparatus consists of a U-tube with an electrode vessel attached near the top of each limb. The right hand limb is supplied with a wide-bore stopcock and the left hand limb or ZE column is open at the top and joined to its electrode vessel via a T-piece. The bottom of the column is fitted with a rubber stopper through which a smooth hole is drilled of such diameter to allow the up and down movement of a capillary tube that passes through it. The application of a little soft grease will ensure that the sliding joint is watertight. The top of the capillary tube is funnel-shaped and the diameter is just less than that of the tube in which it moves and slightly greater than the male ground-glass joint of the electrophoresis column on to which it is ground to form a seal. The sliding funnel serves the purpose of eliminating trailing during fractionation of the ZE column whereby cross-contamination of the fractions is avoided.

Density Gradient Forming Devices

There are a number of density forming devices available in the trade, all of which are suitable for forming density gradients. They are all discussed in detail by Svensson[4]. The device for forming exponential gradients is shown in Fig. 2.

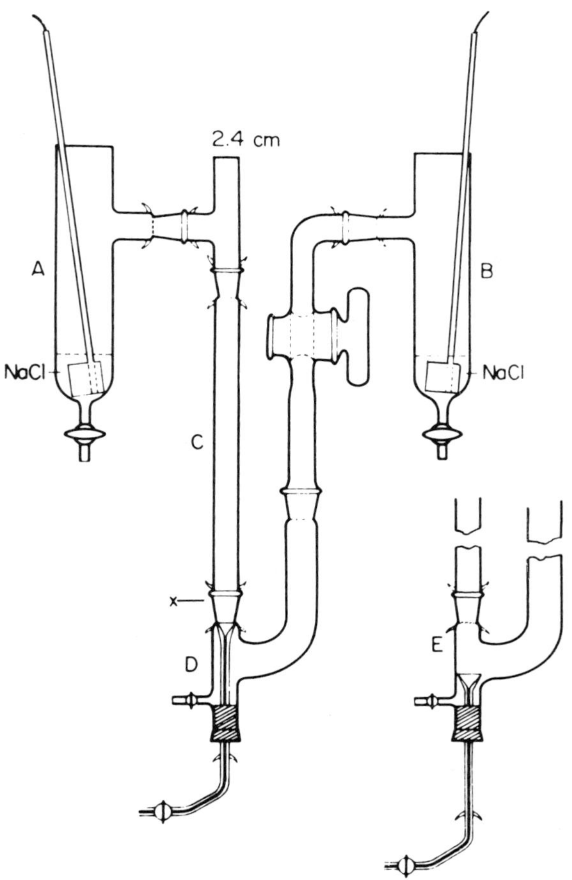

Figure 1. Zone electrophoresis apparatus. A and B are two electrode vessels containing two silver silver chloride electrodes covered with saturated NaCl. C is the column in which the sugar gradient is formed. D is the capillary tube with funnel. The capillary tube slides through the rubber stopper. The funnel is shown in the upper position for formation of the gradient. In E the lower part of the apparatus is shown with the funnel lowered as for electrophoresis. Position X is the approximate level to which the concentrated sugar buffer interface is lowered before the excess sugar solution is run out of the column through the capillary tube.

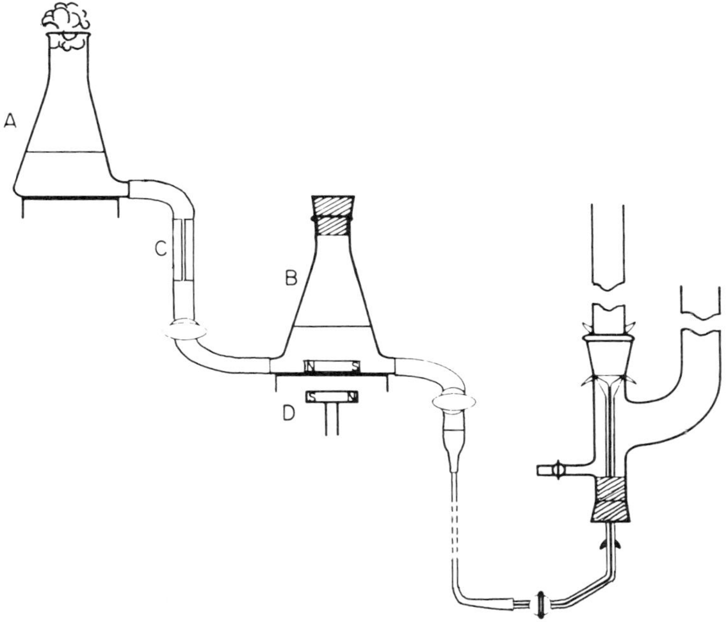

Figure 2. Gradient forming device for formation of exponential-type concentration gradients. A and B are flasks containing the concentrated (40-50%) sugar solution and buffer respectively. C is a capillary and D a magnetic stirrer. A length of rubber tubing of 3 mm bore connects B with the capillary at the lower end of the zone electrophoresis apparatus.

Electrodes

Electrodes used in ZE must be capable of maintaining a constant input of current without the formation of gas bubbles for the duration of an experiment. Gas formation causes electrolytic by-products to be stirred up and mixed with buffer in the electrophoresis tube. Here, changes in the conductivity and pH of the buffer occur. Even if this is avoided by the construction of a special apparatus as in the LKB apparatus, the buffer will be depleted of electrolytes if no additional arrangement is made for continually changing the buffer in the electrode vessels. This complicates the apparatus unnecessarily. Reversible electrodes overcome these difficulties and silver-silver chloride electrodes in saturated sodium chloride have been found to function satisfactorily in ZE.

Silver-silver chloride electrodes may be constructed from two strips of silver gauze 350 x 50 mm each that are welded to two separate lengths of 12 gage pure silver wire. The wires are inserted into glass tubes which are then filled with molten paraffin wax. The silver wire mesh may then be rolled into a convenient shape to fit the electrode vessels. New electrodes must be coated with silver chloride before they are used; this is done by tieing the electrodes together, inserting them in a solution of 0.1 N HC1 saturated with NaCl and connecting them to the positive pole of a source of direct current, with a short length of platinum wire forming the negative electrode. Current is passed until the silver is well coated with silver chloride. When used in the ZE apparatus the electrodes are covered with saturated NaCl in buffer. The silver-silver chloride electrodes will pass a current of 25 m A through the apparatus for 24 hours before gas evolution starts. The electrodes will last almost indefinitely if they are used alternatively at the positive and negative poles. If the duration of an experiment is longer than 24 hours the electrodes may easily be interchanged by switching off the current, closing the wide stopcock and pulling the electrodes from their vessels. After rinsing the electrodes with water, the one with the heaviest (black) coat of AgCl now acts as the positive electrode while the other, the negative.

Buffer

It has been found that the best separation of viruses from associated tissue or other extraneous materials is often obtained at high pH values but within the pH stability region of the virus. It is customary to conduct ZE experiments for virus purification in buffers of pH 8.6. Some of the buffers are:

Borate Buffer

This is prepared as follows: Mix 350 ml of a solution of 62g H_3BO_3 and 20g NaOH in 5 liters distilled water 150 ml 0.1 NHCl, 500 ml 0.15 MNaCl and 1 liter distilled water. The pH of this buffer is 8.6. The molar composition of the buffer is the following:

H_3BO_3	0.035M
NaOH	0.0175M
HCl	0.0075M
NaCl	0.075M

Barbiturate Buffer

This has the following composition:
0.0125 M sodium diethyl barbiturate
0.0125 M diethyl barbituric acid and
0.01 M sodium chloride. The pH of the buffer is 8.6.

Phosphate Buffer

If it is necessary to conduct ZE at lower pH's, 0.015 M KH_2PO_4 and 0.015 M Na_2HPO_4 containing 0.01 M NaCl may be used. By mixing the two phosphate salts in different proportions buffers with different pH's may be obtained (5.9 to 7.9).

Succinic Acid Buffer

A suitable buffer in the pH range 4.2 to 6.0 may be made by mixing 0.025 M succinic acid and 0.025 N NaOH in different ratios.

Acetate Buffer

A suitable buffer for the pH range 4.0 to 5.2 is a mixture of 0.025 M sodium acetate and 0.025 N acetic acid.

Borate buffer is a powerful buffer but it should be borne in mind that when the buffer is used in a sucrose concentration gradient, that the borate ions form complexes with the sucrose resulting in lowering of the pH of the buffer. It is thus necessary to correct the pH by adding N NaOH to the sucrose borate mixture prior to formation of the gradient.

Assembling of ZE Apparatus and Formation of the Concentration Gradient

By studying Fig. 1 it is obvious that the apparatus is held together by a number of Quickfit and Quarts joints. The assembled apparatus on the left of the diagram shows the tiny funnel in position during formation of the density or concentration gradient and during "fractionation" of the column following ZE. The insert on the right shows the position of the tiny funnel during passage of the current.

Initially the apparatus is sterilized with boiling water, the boiling water is run out and replaced with sterile buffer. The electrodes are inserted in the electrode vessels and covered with saturated NaCl in buffer. The concentration gradient forming inert material, is let in through the side arm,

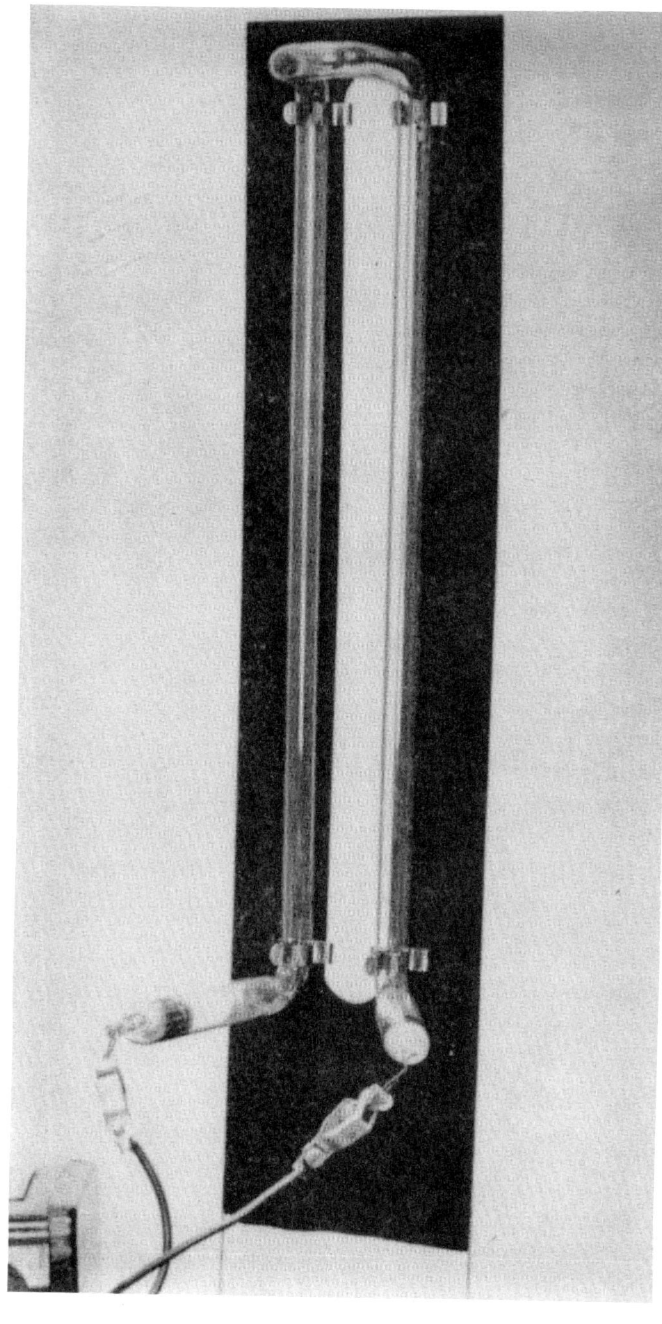

while the funnel is in a position as in the insert on the right of Fig. 1. Sufficient concentrated gradient material, usually 50% sucrose, in buffer is let in to fill the U-bend to a level approximately 2 cm above the "Quickfit" joint. By letting in buffer at the top of the ZE column the 50% sucrose is pushed approximately 2 cm up the right hand limb of the U-tube. The wide stopcock is turned off, the tiny funnel is pushed up against the lower end of the ZE column and any sugar in the funnel drained off. The gradient is now let in through the capillary and funnel from a gradient forming device. The density of the material to be subjected to ZE is adjusted with sucrose to a level which allows the mixture to float on the 50% sucrose in buffer and sinks in the gradient at the lowest level of the gradient column.

The mixture is pushed with a syringe through the capillary to form a layer at the base of the ZE column. To remove the layer from the funnel, 3 ml 50% sucrose in buffer is pushed into the capillary which moves the layer into the column. The funnel is carefully lowered (See Fig. 1 insert) and time is allowed for the sucrose to form a gradient by diffusion across the sample layer. A trace of phenol red in the sample serves as reference to gauge the progress of electrophoresis. The pigment has a constant mobility independent of the pH of the buffer. A second pigment haemoglobin is often used. This protein has a low mobility when compared with that of phenol red. In a sucrose gradient at pH 8.6 the phenol red would migrate 16 cm and haemoglobin 3.0 cm through the gradient. The driving force for this movement would be 3.5 V/cm (230 V across the electrodes). If it is required to extend the ZE period the electrodes may be interchanged.

Fractionation of the Column

At the completion of an experiment the current is switched off and the large stopcock closed. A centimeter scale is attached to the column and buffer is removed from the top until the meniscus is in line with the 0.0

Figure 3. A photograph of the argon - mercury fluorescent lamp used for dark field illumination of the zone electrophoresis column. The lamp consists of a glass tube bent as shown. The tube is attached to a darkened perspex shield with a 2 cm wide slot cut along its length. By placing the slot against the ZE column at the completion of a run the necessary illumination is obtained for observation and photography of light scattering zones in the column.

position on the scale. The meniscus may be used as an index point when fractionating the contents of the column. The funnel is pushed gently with its capillary tube until it is securely on the ground glass joint. The extent of migration of the two pigments, phenol red and haemoglobin is measured and the column inspected for light scattering. If necessary a photographic recording is made. Fractions are now removed from the column. These are usually taken at 1 cm intervals (volume approximately 3.0 ml) and collected in sterile containers. The fractions are then analyzed by appropriate means.

Examination of ZE Columns by Tyndall Light Scattering

The preliminary purification procedures prior to ZE should ensure that the bulk of the soluble proteins present in the crude virus preparation is appreciably reduced. Tissue proteins of approximately the same size as the virus will still be present. These components comprise a considerable proportion of the total material, and are usually visible as light-scattering zones when the column is illuminated by a bright light. Lamps were constructed for the purpose of inspecting and photographing the columns after ZE. A lamp with a fluorescent tube of the same length as the ZE-tube is encased in a metal cylinder with a slot, 1 cm wide cut down the length of the casing. By bringing the slot close to the ZE column the regions of light scattering or opalescence can be identified and photographed at 90° with the slit. A second type is an argon-mercury lamp (Fig. 3) which, because of the short wave length (blue) indicates the positions of the zones of the light scattering components very effectively. After completion of an electrophoresis experiment the lamp was placed parallel to the column and about 1.0 cm away. The positions of the reference substances and zones of light-scattering material were recorded photographically. The photographic negative so obtained may be analyzed in a densitometer thus providing a graphical representation of the components in the ZE column. It was found that this method detected light scattering zones too faint to detect by eye. The degree of purification achieved by ZE may be estimated by comparing the densitometer recording with the infectivity diagram. Such a diagram is shown in Fig. 4.

Ultraviolet-Absorption Recording

The contents of the ZE column may also be pumped through a UV absorptiometer with a peristaltic pump and the effluent collected in a fraction collection. In this manner additional data may be obtained of the

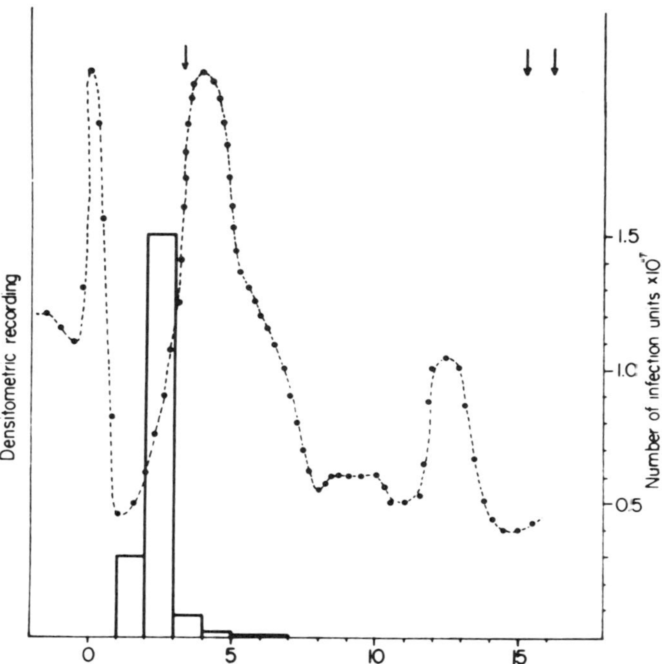

Figure 4. Diagram indicating the position and intensity of light scattering zones in relation to infectivity after ZE of Coxackie A5 virus. The curve of light scattering intensity (dotted line) was obtained by photography using dark-field illumination and subsequent densitometric measurement of the photographic negatives. The histogram indicates the region of virus infectivity. The positions of the reference pigments are indicated. The single arrow represents haemoglobin and the double, phenol red.

material subjected to ZE. A disadvantage of the method is that a certain percentage of the virus may be inactivated by the ultra-violet light of the UV absorptiometer. The percentage inactivation would be an inverse function of the rate of passage of the fluid through the quartz cell of the absorptiometer.

Precautionary Measures

There are several factors which should be observed if correct conclusions are to be drawn from the results of ZE experiments if the technique is to be used for the purification of viruses. When ZE are performed in sucrose gradients it is important that most of the associated tissue material be removed before the sample is inserted in the gradient column. Failing this, the high concentration of sucrose in the column may precipitate the tissue material whereby a considerable amount virus is "trapped" or occluded in the precipitate. As the virus is held mechanically in the precipitate, it may continuously be liberated in the gradient during ZE giving the impression that the virus is inhomogeneous. The phenomenon is particularly noticeable with the large viruses such as those of the pox group, herpes types and influenza.

Importance of Having a Density (Concentration) Gradient Within the Sample Layer

Serious errors may occur when ZE is started immediately after the sample has been introduced in the absence of a gradient across the layer. The components in the mixture will migrate and if there is no supporting gradient, convection will start immediately and will ruin the experiment. It has been found that if the ZE is conducted in a sucrose gradient from a material layer 10 mm wide, 1 hour would be adequate to establish a gradient across the sample.

Optimal Voltage Gradient and Concentration of the Sample

From the results of several experiments at different voltage gradients and current densities it was established that the maximum voltage gradient of 3.5 V/cm and a current density of 18 to 22 mA appear to be optimal, provided a current of air from a fan is blown over the gradient column to remove any heat generated during the ZE experiment.

With the present apparatus the protein concentration should not exceed 20 mg/ml. If this concentration is exceeded the phenomenon known as "raindrops" may occur. This is caused by the large difference in diffusion coefficients between that of the gradient forming material (i.e. sucrose) and the protein mixture. The sugar diffuses readily upwards in the protein layer, resulting in a reduction of the density in the sugar immediately below the protein layer. If the protein concentration in the layer is too high, the protein sugar mixture attains a density greater than that existing in the region from which the sugar diffused. The increased density causes the protein sugar mixture to sink as droplets or streamers.

Partial Purification of Viruses Prior to Zone Electrophoresis

The successful separation of viruses from extraneous matter for ZE depends on initial purification procedures. The most important consequence of using very impure preparations for ZE is that the virus may be transported mechanically by the tissue proteins and cellular debris. The method of purification adopted is dependent on the source and properties of the virus.

Organic Solvents

The picorna viruses which include foot and mouth disease virus and the enteroviruses may be partially purified from infected brain and mouse carcasses by three cycles of ultracentrifugation of the emulsions of the infected material. Using a Beckman preparative centrifuge at 18,000g for 10 min removes coarse tissue debris. Centrifugation at 108,000g for 90 min serves to pellet the virus, and a final cycle at 18,000g for 10 min after resuspending the pellet will remove any large particles present.

A large number of picornaviruses and the majority of the plant viruses are resistant to the organic solvents such as chloroform, chloroform-butanol mixtures and "Freon". These compounds may thus be used to remove further contaminating protein materials prior to ZE of the virus.

Differential Ultracentrifugation and Chromatography

The infectivity of most of the larger viruses is destroyed by the organic compounds mentioned above. They are thus partially purified by differential centrifugation. These viruses usually require albumin to protect them from spontaneous inactivation; serum albumin or ovalbumin must therefore always be included in the suspension medium and in the buffer used for ZE.

If the virus does not lose its infectivity too rapidly in the absence of any one of the albumins the virus may be purified chromatographically on a cellulose ion exchange column or by exclusion chromatography followed by concentration in the ultracentrifuge prior to ZE.

Polymer Displacement

Unstable viruses may also be purified by displacement with mild protein displacement agents such as polyethylene glycol Mr 6000 or polyvinyl

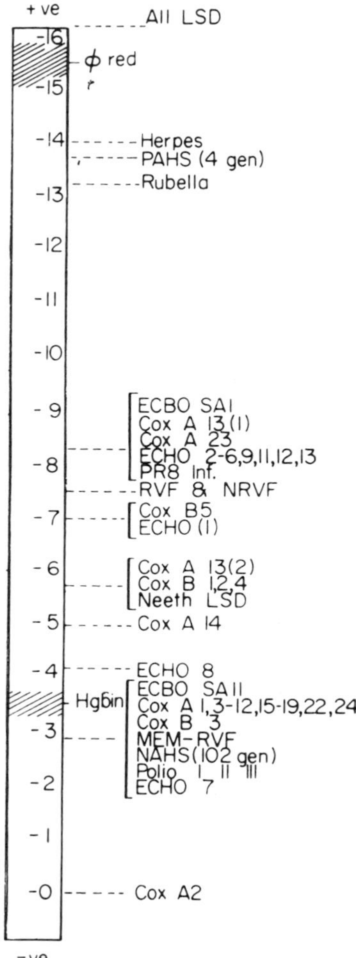

Figure 5. Zone electrophoresis (Composite) diagrams indicating the maxima of virus activity in relation to the distance migrated by rabbit haemoglobin and phenol red (Hgbin and Ø red) in borate buffer at pH 8.6. Animal viruses : All. LSD, Allerton lumpy skin disease of cattle; PAHS, Pantropic African horsesickness virus; RVF and NRVF Pantropic and Neurotropic (105 mouse brain generations) Rift Valley fever virus respectively; Neethling LSD Neethling lumpy skin disease of cattle; MEM-RVF, Rift Valley fever virus that had 105 generations in mouse brains followed by 56 generations in embryonated eggs and 15 generations in mouse brains; ECBO SAI and ECBO II, two bovine orphan viruses; NAHS neurotropic African horsesickness virus (>100 generations in mouse brains, Vaccine strain). The human enteroviruses are indicated by their commonly known denotations.

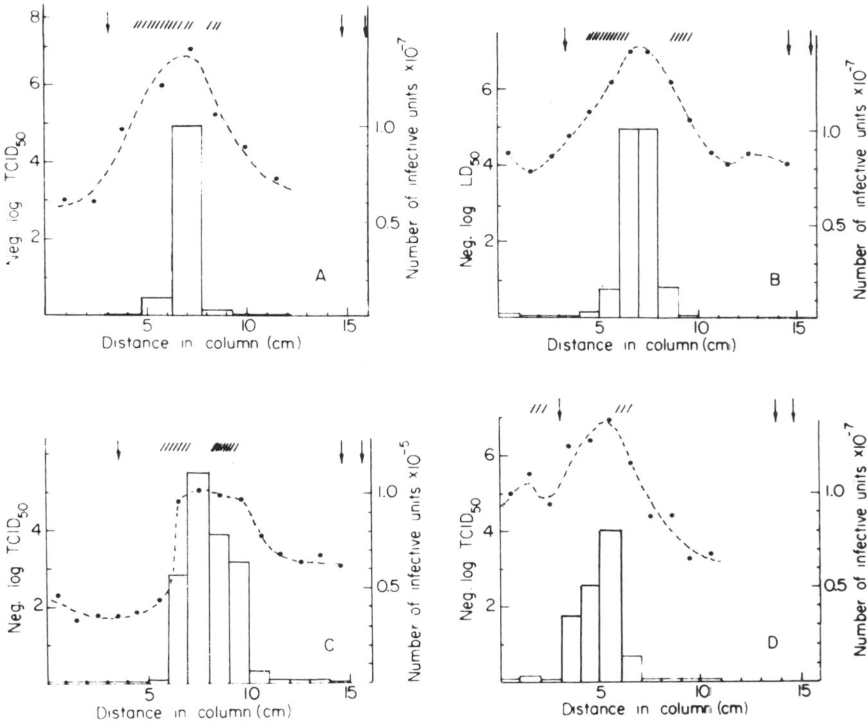

Figure 6. Zone electrophoresis diagrams of RVF suspensions derived from A: infected HeLA cell culture; B: infected mouseblood; C: infected sheep kidney cell culture; D: infected foetal lamb kidney cell culture. The single arrow indicates the position of the haemoglobin and the double arrows the phenol red. Cross hatchings indicate the positions of light scattering zones. For comparison both arithmetic (histograms) and logarithmic (dotted lines) infectivity curves are presented.

pyrilidone. The advantage of using these substances for obtaining fractions rich in viruses is that the polymers are electrically neutral causing them to stay at the position at which the displaced material contaminated with the small amount of polymer was introduced.

(The use of polyethylene glycol for virus purification is described later in this paper).

Haemagglutination

The viruses that haemagglutinate may be purified by adsorption and elution from erythrocytes followed by ultracentrifugation and ZE.

RESULTS

Zone Electrophoresis of Animal Viruses

Enteroviruses

A useful means of expressing the ZE mobility of a virus in an electric field is by means of the RØ factor. The factor equals the distance which the virus migrated, X, from its position of introduction divided by the distance through which phenol red (Ø red), which was added as a trace to the sample initially, migrated (RØ).

The advantage of using Ø red as a standard is that its mobility in the electric field is constant irrespective of the pH of the solution in the gradient column and greater than the majority of viruses. It is advisable to have a second standard pigment but of low mobility. The purpose being that, should any leakage or any displacement of the contents of the ZE column occur, the error could be rectified. A suitable second pigment is haemoglobin.

The ZE experiments were all conducted in sucrose concentration gradients in borate buffer of pH 8.6. The fact that borate ions form complexes with sucrose causing the pH to drop from 8.6 to 7.2 was considered and the pH was readjusted to 8.6 with N NaOH prior to formation of the gradient. For comparing the mobility of one virus with that of another it is essential to have the same experimental conditions whenever possible.

With a few exceptions the majority of human enteroviruses have mobilities, expressed as RØ values, which fall into several groups. The diagram based on their RØ values (Fig. 5) indicates at which levels in the ZE column the different classes of the enteroviruses have their zones relative to haemoglobin (rabbit) and Ø red together with their RØ values calculated from their positions relative to the distance Ø red migrated.

It must be mentioned at this stage that the positions occupied by the zones of the enteroviruses in the ZE column is of the entities freed of extraneous components by chloroform treatment and that the results were reproducible. For instance, the three poliomyelitis types always occupied the same position relative to the rabbit haemoglobin and Ø red. The remaining enteroviruses always had reproducible RØ values in ZE.

Animal Reoviridae (Arboviruses)

The purification of Rift Valley Fever and African Horse Sickness viruses is given in fair detail including problems encountered. The object of this account is to present models for the purification of other insect borne viruses such as West Nile, Yellow fever and Semliki Forest viruses. Results obtained from the study of this group of viruses are interesting but difficult to interpret with certainty. It appears that in ZE a large portion of the infectivity of these viruses is often associated with nonviral opalescent material of the host. In some instances this may be due to similar mobilities of the host material and the virus particles but experiments indicate that there is possibly an association between the host material and the infecting agent. Possible causes are:

a. Coarse host tissue debris tends to precipitate in the concentrated sucrose of the gradient and a high percentage of the virus present may be occluded in the precipitate and be transported mechanically.
b. There is a possibility that the opalescent zone of the host material contains receptor binding material. According to the theory of Baron et al[5] this receptor-site material attaches firmly to the virus and is possibly a lipoprotein as these workers found that its inhibitory properties were lost on treatment with ether.

Examination by isopycnic ultracentrifugation in CsCl of the uninfected host material obtained from ZE columns indicated that it had a density of 1.1 to 1.2g/cc and is thus possibly a lipoprotein with variable protein-lipid ratio. Evidence in favour of the receptor-site theory was obtained when it was found that Rift Valley fever virus (RVF) when grown in different tissues always follows the position of the opalescent zones of the host cellular debris in ZE experiments. The following ZE diagrams of RVF virus (Fig. 6) grown in a variety of hosts are given. It is obvious that the presence of the host debris in the region of maximum infectivity hampers the purification of the virus. After several unsuccessful attempts, success was attained by ZE of pantropic RVF partially purified from the blood of mice infected with the virus. It is evident from Fig. 6B that the maximum titre of the fractions fell almost exactly midway between two opalescent zones of extraneous matter. The virus was separated from the sucrose by centrifugation at 32,000 g for 60 min. The pellet obtained was resuspended in 0.1 M phosphate buffer pH 7.4 and centrifuged once more into a pellet. The electron micrograph, Fig. 7, was made of the virus separated by ZE followed by removal of the sucrose by ultracentrifugation. The virus was negatively stained with phosphotungstate.

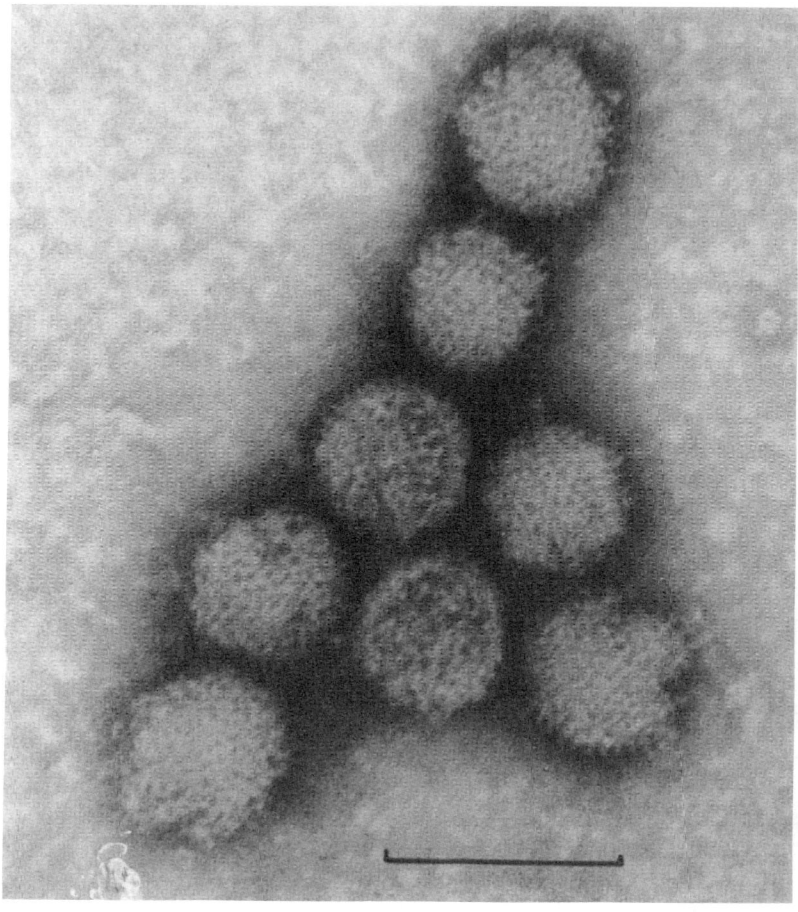

Figure 7. Electron micrograph of pantropic RVF virus obtained from the region between the two zones of light scattering material Fig. 6B. The virus was negatively stained with phosphotungstate; scale line 100 nm, RVF particles 70 nm in diameter.

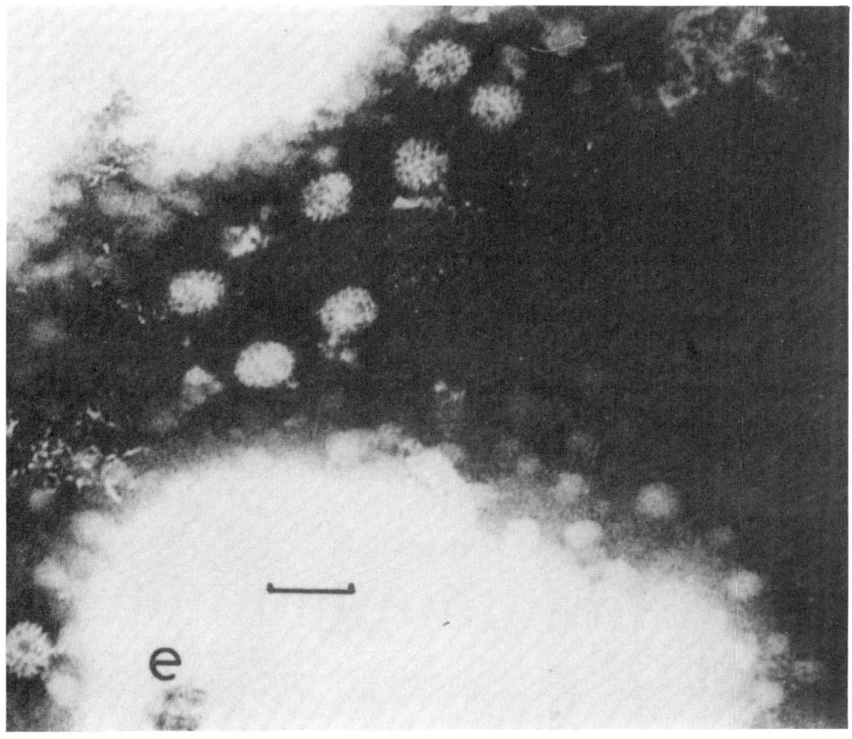

Figure 8. Electron micrograph of neurotropic African horsesickness virus partially purified by displacement with polyethylene glycol Mr6000 followed by a short term zone electrophoresis. The material was negatively stained with phosphotungstate. Particle diameter 70 nm, scale line 100 nm.

African Horsesickness Virus

After many unsuccessful purification attempts of neurotropic African Horsesickness Virus (AHV) by pelleting the virus from clarified infected mouse brain extracts using the Beckman preparative ultracentrifuge, it was decided to use a combination of several methods sequentially. The methods were clarification and pelleting, resuspension of the virus, and displacement from suspension with PEG and ZE. Because of the instability of the virus,

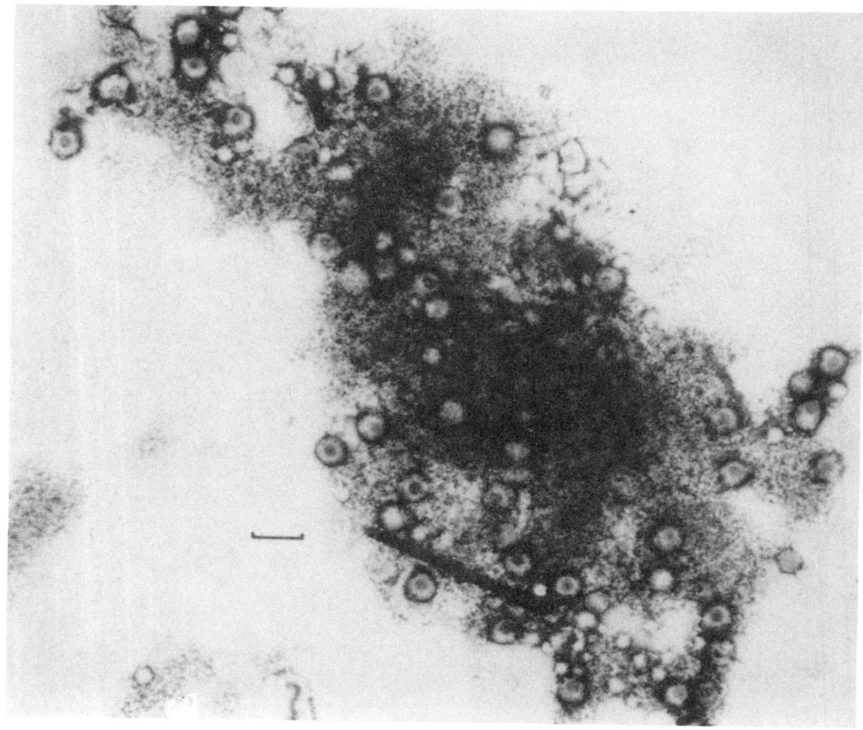

Figure 9. Electron micrograph of All. LSD virus neg. stained with phosphotungstate. Particle diameter 138 nM, line 200 nm.

the minimum time was allotted to each step including the ZE. From a knowledge of the R Ø value of the virus, the position of the virus zone could be determined accurately. The corresponding fraction was "washed" free of the sucrose and subjected to electron microscopy (Fig. 8).

Other animal and human viruses which have been investigated by ZE were lumpy skin disease (LSD) of cattle, Herpes, Rubella and pantropic African horse sickness virus. It is of interest to note that unadapted pantropic horse sickness virus had a high mobility RØ = 0.8 and when the virus was fully adapted by intracerebral injection of mice (the vaccine strain) the virus showed an RØ of 0.2

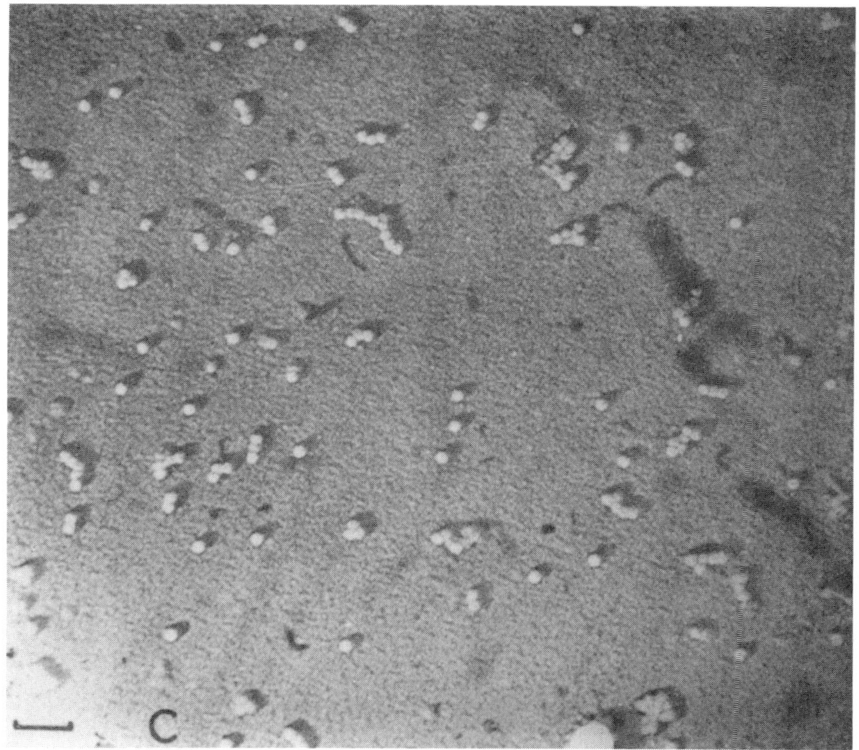

Figure 10. Electron micrograph of ECBO SAI virus shadowed with WoO_3. The polystyrene particle 138 nm; virus particles 29 nm; line 200 nm. Virus purified by zone electrophoresis.

Additional electron micrographs of some of the human and animal viruses purified by ZE are given in Fig. 9, 10, 11.

Zone Electrophoresis of Plant Viruses

The application of ZE to plant viruses has been rewarding, the reason being that these viruses are generally available in large amounts in infected plant tissue and in relatively pure states compared to animal viruses. Not only has ZE proved effective in producing pure preparations but it has also

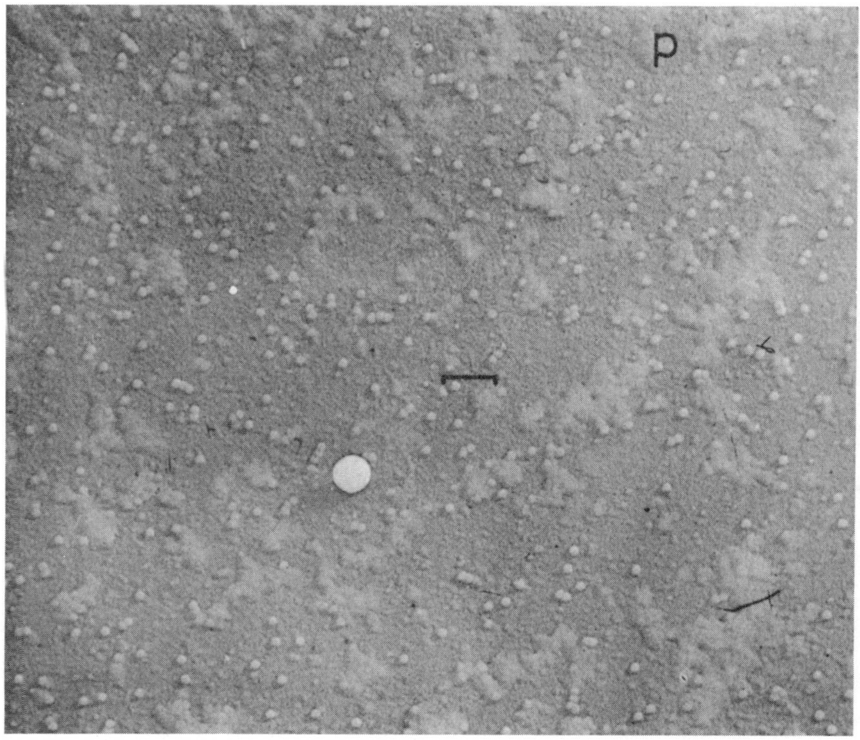

Figure 11. Electron micrograph of Coxsackie B3 virus purified by zone electrophoresis in a sugar concentration gradient. Amorphous material amongst the virus particles is very likely remaining sugar not removed by washing; shadowed with W_o0_3; particle diameter 29 nm; line 200 nm.

assisted in the identification of the infective agents. For instance, Van Regenmortel et al.[6,7,8] have found that pathological conditions in plants may be caused by more than one virus. These researchers have isolated three different viruses from a single plant that showed symptoms of Calico disease.

Unlike animal viruses of particle diameters >50 nm, plant viruses are resistant to chloroform chloroform-butanol mixture or other organic

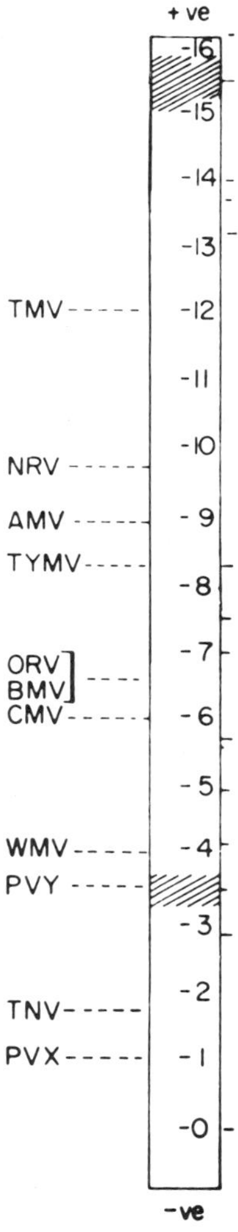

Figure 12. Composite diagram of plant viruses indicating the distances migrated in zone electrophoresis relative to the distances migrated by the reference pigments haemoglobin (Hgbin) and Ø red: TMV tobacco mosaic virus, NRV necrotic ringspot virus, AMV alfalfa mosaic virus, TYMV turnip yellow mosaic virus, ORV, BMV and CMV, odontoglossum ringspot, bean mosaic and cucumber mosaic viruses respectively; WMV, watermelon mosaic virus, PVY and PVX potatoviruses respectively and TNV tobacco necrosis virus.

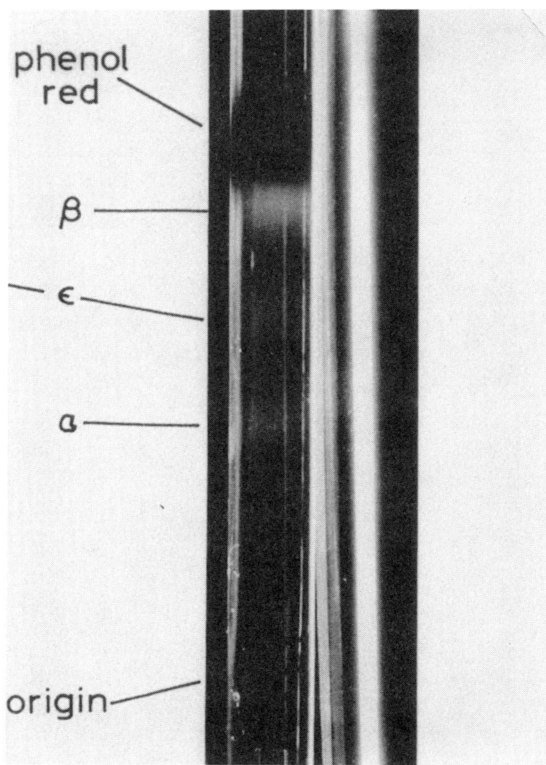

Figure 13. Zone electrophoresis column showing the zones of three viruses separated from an extract of moribund *Nudaurelia cytheria cytheria*. The intense zone below the Ø red is that of the Nβ virus which is the predominant infective agent of Nudaurelia caterpillars. Two faint zones of Nudaurelia ϵ and α are just visible below the main zone.

solvents; infected plant extracts may be deproteinized with these solvents leaving the plant viruses unaffected. The damaging effects which are often noticed when pellets obtained by ultracentrifugation of the watery phase containing the virus following chloroform-butanol treatment, for example seen on electron micrographs, are due to tight packing in pellets and redispersion with Pasteur pipettes. The flexuous viruses PVX and PVY° and those belonging to the tobamo group are particularly vulnerable to

Table 1. RØ values of the N viruses calculated from the distances the viruses migrated relative to that migrated by phenol red in a sugar gradient at pH 8.6.

Nudaurelia Virus	RØ values
Nα	0.5
Nβ	0.93
Nγ	0.16
Nδ	0.33
Nϵ	0.67

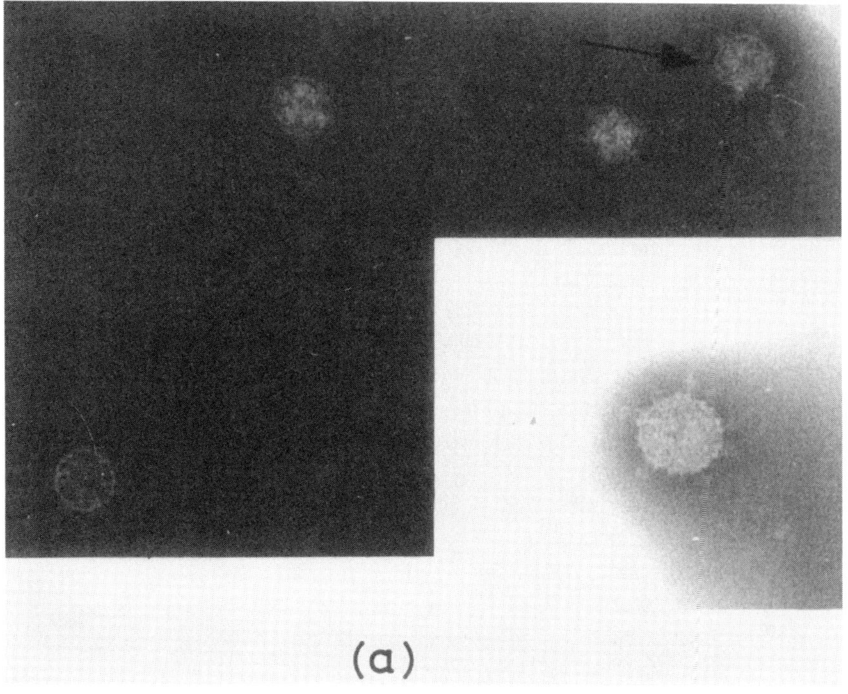

Figure 14. Nα virus separated by zone electrophoresis. The virus seems to disintegrate on the electron microscope grid hence its apparent low concentration. The insert indicates that the virus particle has a number of spherical units on its surface; phosphotungstate staining Magnification 200,000 fold.

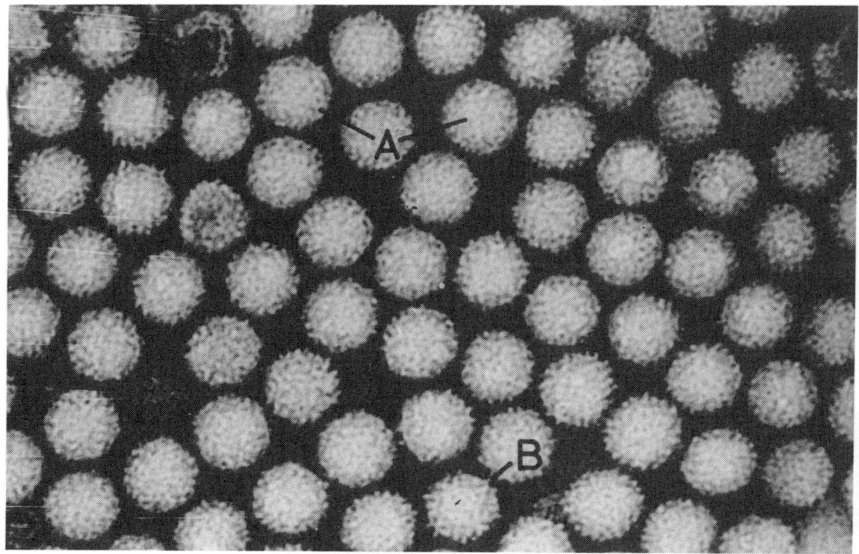

Figure 15. N β virus separated by zone electrophoresis, negative staining, phosphotungstate. Magnification 230,000 fold.

damage during redispersion. (To avoid breakage of these viruses the technique of the thin layer ultracentrifugation was introduced - see details in this communication).

Difficulties are sometimes encountered with the fraction 1 antigen (F1). This antigen occurs in a large number of plants and if an anti serum is to be elicited in a suitable animal against a plant virus, it is important to free the virus antigen completely of the F1 component[6,7,8] as this protein is a potent antigen. Van Regenmortel eliminated this antigen by ZE or if the virus has a mobility similar to that of the F1 antigen to grow the virus in a plant which has an F1 antigen with a different electrophoretic mobility. An immunological method used by the author (unpublished work) appeared to be effective for removing the F1 by precipitation prior to ZE. Other methods for removing the F1 are gel exclusion chromatography and ion

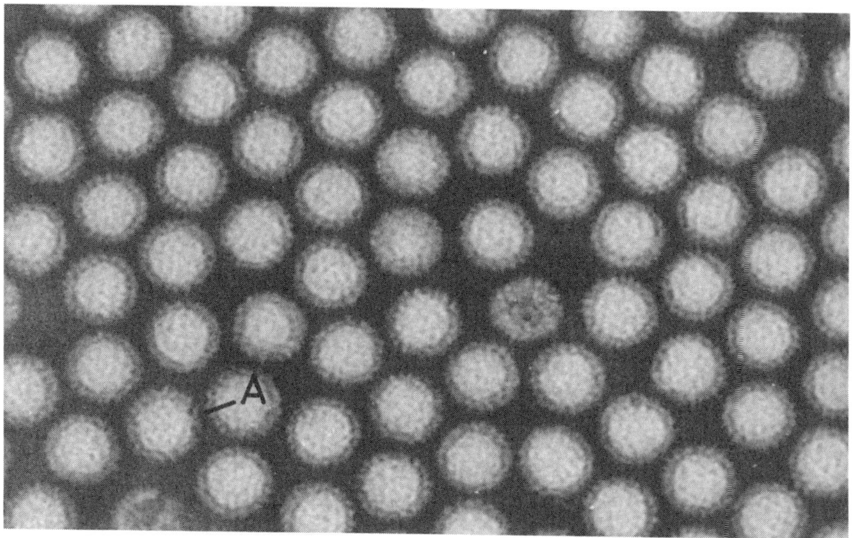

Figure 16. N ϵ virus separated by zone electrophoresis negative stained with phosphotungstate. Magnification 300,000 fold.

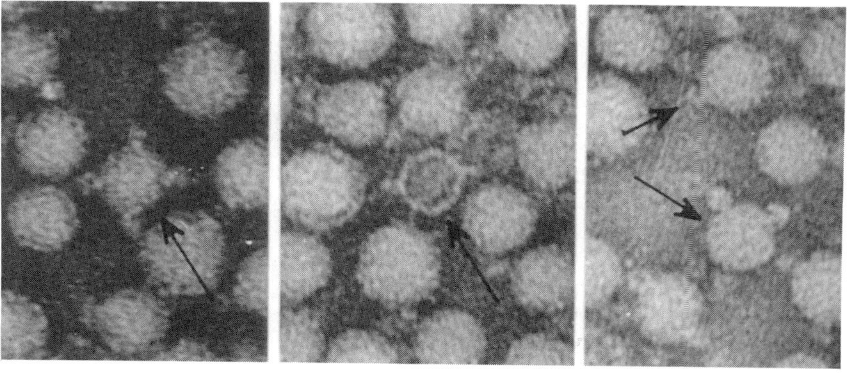

Figure 17. N δ virus separate by ion exchanging (Ecteola). Magnification 550,000 fold. Negative stained with phosphotungstate. Preparation contaminated with Nγ. The Nδ particles show spherical bodies on their surfaces. Arrows indicate Nδ particles.

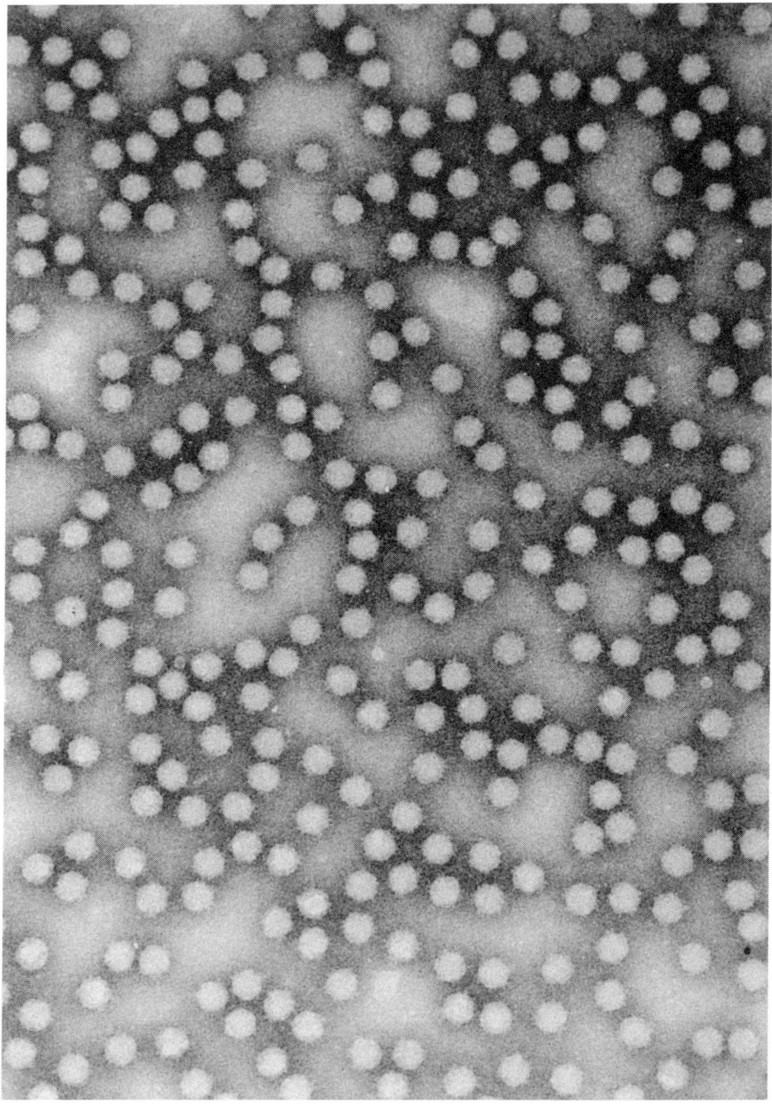

Figure 18. Nγ zone electrophoresis, negative staining phosphotungstate. Magnification 200,000 fold.

exchanging chromatography. Gel exclusion is effective for separating spherical virus particles from F1 but the method is less suitable for separating filamentous virus particles from the extraneous antigen because of trailing of the virus filaments.

In the composite diagram Fig. 12, the positions of maximum activities of the plant virus zones relative to Ø red and rabbit haemoglobin in a ZE column are given.

Insect Viruses

Permission was obtained from Dr. I. R. M. Juckes and the University of Cape Town for the incorporation of some data, including electron micrographs of the viruses of the Pine Emperor Moths, *Nudaurelia cytheria cytheria* (N) in this communication. The data was taken from the Ph.D. thesis of Juckes[9].

Juckes isolated the viruses from sick and dead N caterpillars using a series of techniques which involved disintegration in phosphate buffer in a Waring blender, clarification of the pulp by centrifugation at 1,000 g for 30 min and deproteinizing by vigorous shaking with chloroform-butanol 9:1 mixture. The ensuing emulsion was centrifuged at 1,000 g for 10 min, the watery phase, which contained the viruses separated and the infective agents displaced with PEG of increasing concentrations. Finally the displaced fractions were subjected to zone electrophoresis (ZE) in concentration gradients of sucrose. Juckes established that the different N viruses had RØ values differing sufficiently from one another to enable him to separate them, with one exception, uncontaminated from one another. He labelled the viruses N(α, β, γ, δ, ϵ). Of the viruses Nβ was the most abundant followed by Nϵ and Nγ. The Nα and Nσ virus occurred in trace amounts in the extracts of the caterpillars. It was nevertheless possible to isolate and partially purify the two entities and have their morphologies determined by electron microscopy.

In Table 1 the different viruses and their respective RØ values are given.

In Fig. 14 to 18 electron micrographs are given of the different N viruses in the extracts of the N caterpillars.

REFERENCES

1. Svensson, H. and Valmet, F. Science Tools, *2*, 11 (1955).
2. Cramer, R., Lerner, K. D. and Polson, A. Science Tools *4*, 1 (1957).
3. Polson, A. and Cramer, R. Biochim. Biophys Acta *29*, 187–192 (1959).

4. Svensson, H. In laboratory Manual of Analytical Methods of Protein Chemistry (P. Alexander and R. Block eds. Vol 1, pp 193–244. Pergamon Press. Oxford (1960).
5. Baron, S., Friedman, R.M. and Buckler, C.E. Proc. Soc. Exp. Biol. Med. *113*, 107–110 (1963).
6. Van Regenmortel, M.H.V. Virology *54*, 282 (1964 a).
7. Van Regenmortel, M.H.V. Virology *23*, 495 (1964 b).
8. Van Regenmortel, M.H.V., Nel, A.C. and Hahn, J.S. S. Afr. J. Agr. Sc. *7*, 817 (1964).
9. Juckes, I.R.M. PhD Thesis, University of Cape Town (1973).

2. A Theory for the Displacement of Proteins and Viruses with Polyethylene Glycol*

ABSTRACT

The displacement action of polyethylene glycol of different molecular weights may be linked to the ability of the polymers to form coiled particles in solution. From conclusions drawn from their sedimentating properties in centrifugal fields the polyethylene glycols of low molecular weights, as expected, are less randomly coiled than those of higher molecular weight. It is suggested that protein molecules have the ability to diffuse into the coils of the polyethylene glycol from which they are excluded when the random coiling increases with increasing polymer concentration. From considerations based on the interaction of the polymer filament with the displaced particle the distribution of the substance between the coils and the intermolecular spaces may be predicted semi-quantitatively.

INTRODUCTION

Until recently the mechanism whereby a macromolecular substance or virus can be precipitated from solution by polyethylene (PEG) glycol remained obscure. Several attempts have been made to devise a workable hypothesis [1 2,3,4,5,6,7,8], but none of these could satisfactorily explain all the

*From: A. Polson, Prep. Biochem. 7:129-154 (1977).

phenomena connected with the precipitation. It is generally believed that the precipitation is a displacement reaction of one colloid by another.

This work proposed a new theory which is based on conclusions drawn from the sedimentation behaviour of polyethylene glycol (PEG) fractions of widely differing molecular weights and on similar data on methyl-cellulose fractions also of differing molecular weights.

MATERIALS AND METHODS

Polyethylene Glycol

Polyethylene glycols of molecular masses weights 1,500, 6,000 and 20,000 Daltons (D) were obtained from Shell Chemicals, Cape Town. In preliminary diffusion measurements the 1,500 and 6,000 D varieties appeared to be homogeneous but the 20,000 D heterogeneous. The latter was subjected to fractionation with ammonium sulphate and yielded a fraction which appeared to be homogeneous and which had a molecular weight of approximately 20,000 D by intrinsic viscosity.

Methyl-Cellulose

Ultracentrifugation and diffusion data were obtained by Signer and von Tafel[9] and by Polson[10] respectively, on the same fractions of methyl-cellulose. From their behaviour during diffusion it was established that they were homogeneous and their molecular weights were of the same order of magnitude as those of PEG fractions under investigation.

Influenza Virus

The x31 strain of influenza virus was inoculated into embryonated eggs and the infected allantoic fluid harvested after three days incubation at 37°C. Virus concentration was measured by haemagglutination (HA) units per unit volume.

Albumin

Bovine albumin, supplied by Miles Ltd., Cape Town was of a purity in excess of 98%. A 1% solution in phosphate buffer pH 8.0 was divided into a number of aliquots. To these was added pulverized PEG Mr 6,000 D and the albumin precipitates which formed were centrifuged off at 15,000 x g.

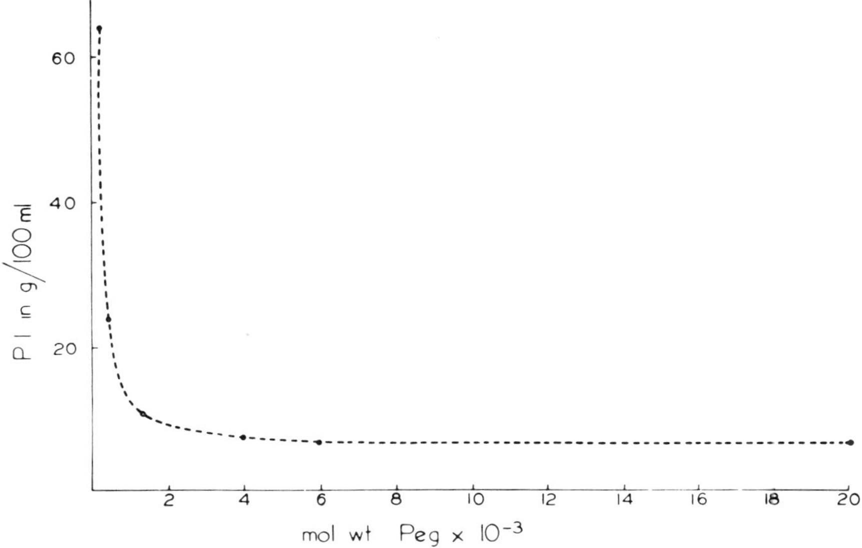

Figure 1. Effect of molecular weight of PEG on its displacement index (% PEG necessary to displace 50% gamma-globulin from a 1% solution).

The u.v. adsorption at 280 nm was measured for each SNF and the results plotted after correction for the increase of volume following addition of the polymer.

Gamma-Globulin

The results on precipitation of gamma-globulin with PEG of a wide range of molecular weights were taken from a previous publication[11]. The precipitation indices, taken as the concentration of the polymer which caused the precipitation of 50% of the gamma-globulin from a 1% solution, are plotted against the molecular weight of the polymer. The data are shown in Fig. 1.

Table I. Sedimentation Coefficients and Molecular Weights of Polyethylene Glycol Fractions

PEG D	S_{20W}
1,500	0.0
6,000	0.49
20,000	0.8

Table II. Sedimentation Coefficients and Molecular Weights of Methyl-cellulose Fractions

Methyl-cellulose D	S_{20W}
14,200	0.83
22,600	0.79
33,000	0.89

Analytical Ultracentrifugation of Polyethylene Glycol

The PEG fractions of Mr 1,500, 6,000 and 20,000 D were ultracentrifuged in the Model E at 56,100 rpm and 20°C. The dispersion medium was water. Exposures were made at 64 min intervals and the concentration of the solution was 0.2 g/100 ml.

RESULTS

In Tables I and II are the sedimentation results obtained with PEG and methyl-cellulose fractions respectively. The molecular mass shown for the 1,500 and 6,000 D PEG entities were those given by the manufacturer but the value for the 20,000 D material was calculated. The molecular weights of the methyl-cellulose fractions were calculated from their sedimentation, diffusion and partial specific volume data by the well known Svedberg equation.

Precipitation Attempts of Gamma-Globulin with Methyl-Cellulose

Attempts to precipitate gamma-globulin at 1% concentration with unfractionated methyl-cellulose were unsuccessful due to the extremely high

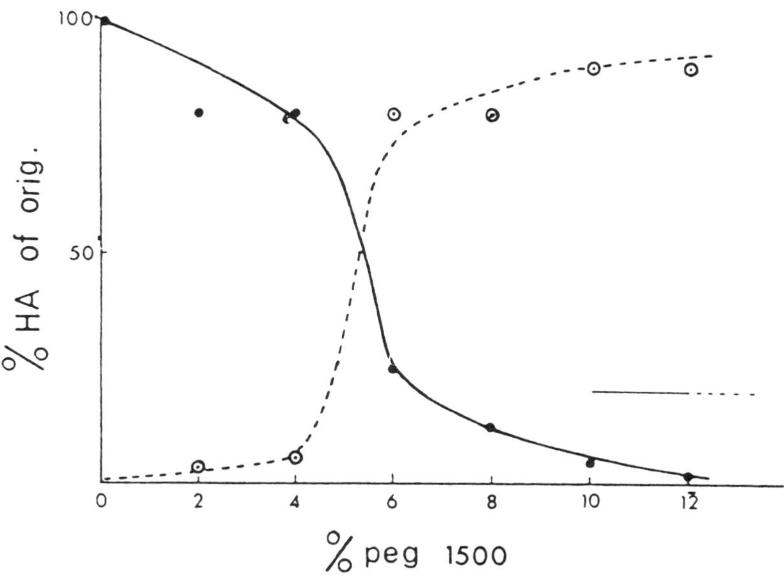

Figure 2. Displacement of the x31 recombinant strain of Influenza virus from infected chicken allantoic fluid with PEG 1,500 D. The ordinates are residual percentage haemagglutinin in the SNF after treatment with PEG 1,500 D of different concentration. Continuous curve = residual HA in SNF; interrupted curve = redispersed HA precipitate. Horizontal line indicates concentration of PEG where proteins precipitated.

viscosities of methyl-cellulose solutions at moderately low concentrations (2-4%). No precipitation occurred in this concentration range.

Precipitation of Influenza Virus with PEG

The results obtained with precipitation of aliquots of influenza virus with PEG of different molecular weights are presented in Figs. 2, 3 and 4. The data shown for PEG Mr 6,000 D were taken from a previous publication[12]. To enable a comparison to be made of the effectiveness of each of the three types of PEG in precipitating influenza virus, the concentration of PEG necessary for the removal of 90% of the virus activity (HA) from the suspensions was used as an index. These results are presented in Table III.

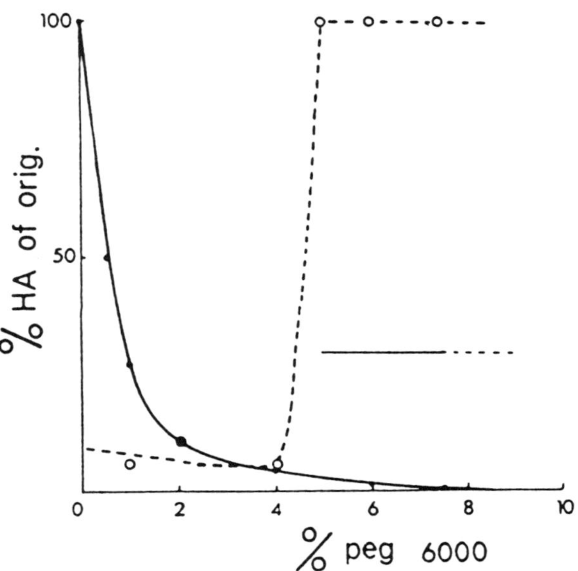

Figure 3. Displacement of x31 strain of Influenza virus from infected allantoic fluid with PEG 6,000 D. Ordinates, residual haemagglutinin, abscissae, concentration PEG 6,000 D. Continuous curve = residual HA in SNF; interrupted curve = redispersed HA precipitate. Horizontal line indicates concentration of PEG where proteins precipitated.

Precipitation of Bovine Albumin with PEG 6,000 D

The results of precipitation experiments on bovine albumin with PEG 6,000 D are presented graphically in Fig. 5.

Shape and Configuration of Polyethylene Glycol in Solution

From Table II it is evident that the sedimentation coefficient of the methyl-cellulose fractions examined did not vary with the molecular weight. The fluctuations in the values were probably due to the fact that an accurate determination of S values for substances with very low sedimentation coefficients is inherently difficult. The explanation of the phenomenon offered by Signer and von Tafel is very likely to be correct. These authors

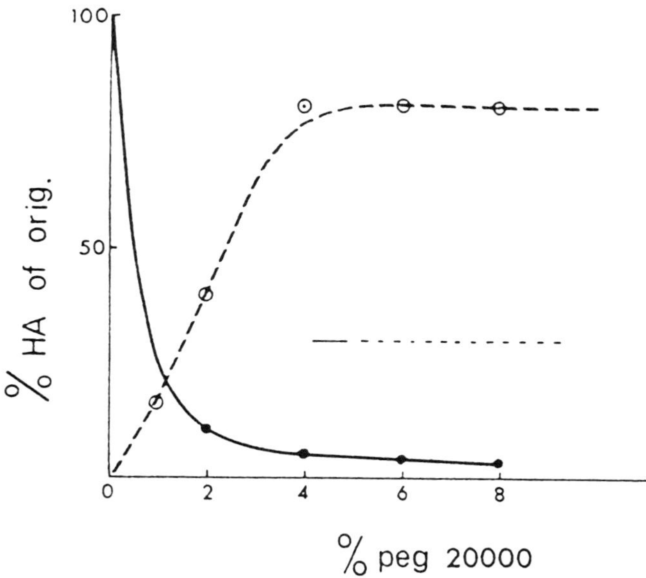

Figure 4. Displacement of x31 strain of Influenza from infected allantoic fluid with PEG 20,000 D. Ordinates, residual haemagglutinin abscissae, concentration PEG 20,000 D. Continuous curve = residual HA in SNF; interrupted curve = redispersed HA precipitate. Horizontal line indicates concentration of PEG where proteins precipitated.

Table III. Precipitation Indices of PEG of Different Molecular Masses for Influenza Virus x31

PEG D	Precipitation Index (% PEG)
1,500	8.5
6,000	2.0
20,000	1.8

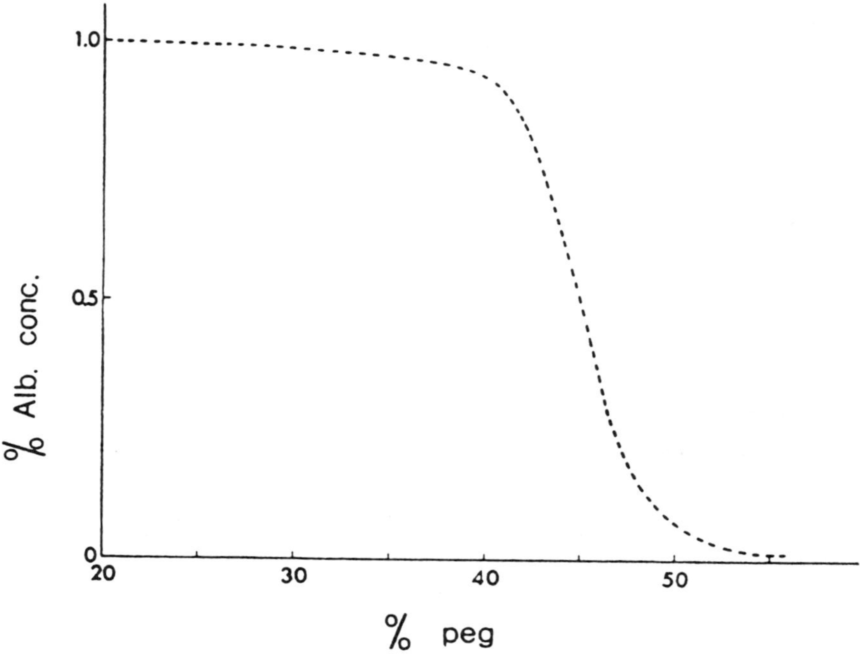

Figure 5. Displacement of bovine albumin with PEG 6,000 D. Ordinates percentage albumin in SNF, abscissae, concentration PEG 6,000. The stock albumin solution was 1%.

suggested that the centrifugal force acting upon the methyl-cellulose is proportional to the length of the methyl-cellulose molecule (equivalent to the molecular weight). The opposing frictional force is also proportional to the length, consequently constancy of sedimentation coefficients with molecular length may be expected.

This implied that methyl-cellulose in water is composed of rod-shaped particles that are rigid or flexible; in their most distorted state they are at least thoroughly drained by the medium (föllig durchspülte Faden[9].).

From the results presented in Table I it is obvious that the sedimentation coefficients of the different PEG entities increase with increasing molecular weight. This implies that the configuration of PEG in a watery solution is different from that of methyl-cellulose. The PEG particles cannot be thoroughly drained linear or flexible rods as the different polymer entities would then have similar sedimentation coefficients. The most fitting explanation for the increase of sedimentation coefficient with increase of molecular weight is that PEG in watery solutions is composed of randomly coiled filaments. Water is trapped in the coiled structure and the folded filament with its encased water behaves like a large heavily hydrated molecule of low density. Increase in molecular weight or particle length results in more intricate coiling and therefore proportionally more water becomes trapped. As the sedimentation coefficient of a particle is proportional to the second power of its diameter it is to be expected that the sedimentation coefficients will increase with molecular weight or chain length of the polymer.

The PEG entity of molecular mass 1,500 D showed no measurable sedimentation during centrifugation, which implies that very little water, if any, is trapped in the molecule and that it is very likely thoroughly drained for the above reason and on account of its small size.

This configuration suggested for PEG is in agreement with Flory's[13] concept of the spatial arrangement of randomly coiled polymer filaments. Flory stated that the configuration which a linear polymer chain in a dilute solution would assume in a good solvent would depend upon the restricting angles between links of the polymer chain. With complete freedom of folding the polymer molecule would resemble the path of a particle in Brownian motion or that of a diffusing gas molecule. In all dilute solutions of real linear polymer molecules the direction assumed by an intramolecular bond will be strongly influenced by that of the preceding bond in the chain. The net effect of randomly coiling in a real polymer would be a more expanded configuration in comparison with that of a freely jointed chain of the same length, although the spatial arrangement would still resemble the path of a particle in Brownian motion.

In dilute solutions of linear polymers the configuration of the molecule is not influenced by similar molecules in its immediate vicinity. In concentrated solutions, however, the effect of identical molecules in its neighbourhood must be taken into consideration. Because of the proximity of similar molecules the spatial extension of the molecule becomes restricted and its Brownian motion of rotation round its center of mass occurs in a smaller space. In the absence of strong intermolecular attractive forces the polymer molecule will assume an approximate spherical configuration with its compactness determined by the restricting angles between successive bonds in the chain, their lengths and proximity of neighbouring polymer molecules.

Displacement of Colloids with PEG

Them more recent theories of linear polymer solutions are mainly extensions of those elaborated by Flory[13] and by Huggins[14,15]. In broad terms these extensions explain in qualitative manner the phase separation which occurs when two incompatible colloids are mixed in relation to the chemical potentials and molecular species involved. The theories of special interest to the present work are those of Edmond and Ogston[7] and of Juckes[8]. Juckes showed by means of new formulations, based on Edmond and Ogston's[7] work and supported by solubility measurements of proteins and viruses in solutions of PEG, that the displacement of these proteinaceous substances occurs according to the "salting-out" principle. Adopting this concept of Juckes and considering that "salting-out" is essentially the deprivation of proteinaceous and other colloidal substances of water through hydration of salt ions, it was concluded that displacement of these high molecular weight substances from solution by PEG is also a result of dehydration. As the removal of water by the polymer cannot be ascribed to binding onto ions a different mechanism was considered.

According to Flory[16] a randomly coiled polymer in solution may be regarded as an osmotic system on a submicroscopic scale capable of swelling and contracting; closely paralleling that which exists in a three-dimensional network of gel or tissue. The osmotic action of the solvent swells the molecule to the extent where the retractive forces of the bonds in the polymer filament come into equilibrium with the osmotic forces which tend to expand the molecule. In accordance with the Flory theory and sedimentation behaviour in the analytical ultracentrifuge, imaginary models of the three PEG molecular species, as they would appear in moderately concentrated solution, are presented in Fig. 6. It was endeavoured to present the three polymer types by folded lines of lengths which are proportional to their respective molecular weights.

Consider a dilute solution of a substance of molecular weight M in the presence of any one of the three PEG molecular types. The molecular diameter of the substance must be small enough to enable it to diffuse to a limited degree into the coils of the PEG molecule and that the submicroscopic system (substance - PEG) is in equilibrium. The condition for equilibrium requires that the osmotic pressure of the substance exterior

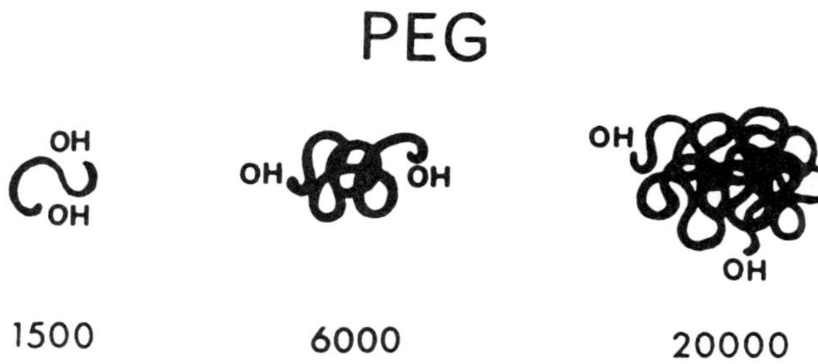

Figure 6. Imaginary models of PEG of molecular masses 1,500; 6,000 and 20,000 D in watery solution. The lengths of the folded chains are proportional to the molecular weights. The hydroxyl groups at the ends of the polymer chains are indicated by OH.

of the coils π_ϵ, should be equal to that in the interior of the coils π_i. As the solution of the substance is dilute,

$$\pi_\epsilon = \frac{RTC_\epsilon}{M}$$

where R, T and C_ϵ are the gas constant, the absolute temperature and the external concentration respectively. The second virial coefficient is omitted on account of the low concentration of the substance. The osmotic pressure of the substance in the coiled structure is

$$\pi_i = \frac{RTC_i}{M} + F$$

in which C_i is the internal concentration and F is a factor related to the second virial coefficient resulting from interaction between the substance at concentration C_i and the coils of the PEG molecule. The factor contributed by the coils is a function of the molecular weight and the concentration of the PEG. Equating this factor with C_pP, in which C_p is the polymer concentration it follows that

$$\frac{RTC_\epsilon}{M} = \frac{RTC_i}{M} + RTC_i\, C_pP$$

Dividing throughout by $\frac{RTC_i}{M}$ follows

$$\frac{C_\epsilon}{C_i} + 1 + MC_pP$$

or

$$\frac{C_\epsilon}{C_i} - 1 = MC_pP \qquad (1)$$

Regarding $\frac{C_\epsilon}{C_i} - 1$ in equation (1) as a qualitative measure of the displacement from the coiled structure of the PEG it is evident that for a given PEG concentration and type, the displacement will be proportional to the molecular weight of the displaced substance, being low for low molecular substances and proportionally higher for substances of higher molecular weight. Equation (1) further indicates that the displacement of the substance from the coils of the polymer is proportional to the concentration of the polymer and to a function P of the molecular weight of the polymer. For substances of very high molecular weight such as the giant haemocyanins or viruses which can only diffuse into the coils to a limited degree, $\frac{C_\epsilon}{C_i} - 1$ will always be large.

Precipitation of the Displaced Substances

Equation (1) describes in a qualitative manner how a substance is preferentially concentrated in the interpolymer molecular spaces. It does not indicate how the substance is precipitated from solution. The model of Flory[16] of a randomly folded linear polymer, regarded as a submicroscopic osmotic system, again provides an answer.

Consider a solution of a substance of high molecular weight such as a protein in osmotic equilibrium with a watery solution of a linear polymer such as PEG and stir into the mixture more polymer in pulverized form. The additional PEG will cause contraction of the polymer coils and displacement of more protein from the coils. The added polymer molecules will also encase solvent in their folds thereby increasing the concentration of the protein in the inter PEG molecular spaces. Eventually through the addition of sufficient polymer the concentration of the protein C_ϵ exceeds its solubility and precipitation commences. It stands to reason that precipitation will start at a lower polymer concentration when the initial protein concentration is high and that dilute suspensions will start to precipitate when the concentration of the polymer is high. The precipitation curve of bovine serum albumin in Fig. 1 illustrates this. The molecular weight of the protein or, more precisely, the Stokes's radius of the protein is of paramount importance in determining its rate of precipitation. This was demonstrated convincingly by Juckes[8]; the larger the Stokes's radius the greater the precipitation rate.

In the case of viruses, on account of their large Stokes's radii, complete precipitation occurs over a narrow range once precipitation has commenced. According to the present theory it is tempting to assume that virus particles, on account of their relatively large sizes when compared to the sixes of protein molecules, do not enter the spaces in the polymer coils.

Parallel Phenomenon of Gel Exclusion Chromatography

It is not possible to test equation (1) quantitatively and it must therefore be regarded as a qualitative description of the factors involved. The related phenomenon of gel exclusion which is widely used in chromatographic separation of proteins and viruses, provides indirect proof of the acceptability of equation (1).

A quantitative theory for this phenomenon based on thermodynamic principles has already been presented by Polson and Katz[17]. For the sake of comparison and clarity the reasoning is again given.

Consider a dilute solution of a protein in contact with a porous gel particle. At equilibrium the protein in the medium and that in the particle will be distributed in such a manner that their respective osmotic pressures are equal, that is:

$$\pi_\epsilon = \pi_i$$

At high protein concentration

$$\pi_\epsilon = \frac{RTC}{M} + RTC_\epsilon^2 B \qquad (2)$$

in which B is the second virial coefficient which signifies the molecular interaction at higher concentration. The other symbols have their previous meaning. The factor B has a low value for compact globular proteins but is appreciably larger for rod-shaped or ellipsoidal particles (Tanford)[18].

For the sake of simplicity assume that we are dealing with a dilute solution of a globular protein in contact with a gel particle. Under these conditions the virial factor is ignored and equation (2) becomes:

$$\pi_\epsilon = \frac{RTC_\epsilon}{M}$$

The partial osmotic pressure inside the gel particle equals

$$\pi_i = \frac{RTC_i}{M} + F \qquad (3)$$

This equation indicates that because the protein particles are now in a confined space in the gel matrix their Brownian movement is now interfered with by the gel fibrils. F is a compounds function of the interference between the protein particles in the spaces in the gel and the gel fibrils. The second virial coefficient of the protein alone may be disregarded but that pertaining to the interaction between protein and the gel fibrils must be considered.

The factor F in equation (3) may be written as

$$F = C_i\ C_g\ G$$

$$\pi_i = \frac{RTC_i}{M} + RTC_i\ G$$

Equating this with π_ϵ and dividing throughout by $\frac{RTC_i}{M}$ follows

$$\frac{C_\epsilon}{C_i} = 1 + MG$$

or

$$G = \left(\frac{1}{K_d} - 1\right)\frac{1}{M} \qquad (4)$$

K_d is the distribution coefficient $\frac{C_i}{C_\epsilon}$

Table IV. Determination of G for 10% Agarose Pearls (100 Mesh)

The elution volume was 66 m M-phosphate buffer pH 7.1, containing 72 m M NaCl

Protein	M(Lit)	K_d	G x 10^5
Cytochrome C	13,000	0.695	3.37
Lactalbumin	16,400	0.661	3.13
Chymotrypsinogen	25,000	0.548	3.30
Ovalbumin	45,000	0.415	3.31
Bovine serum albumin	68,000	0.297	3.48

The factor K_d is determined by the void and elution volumes in gel exclusion chromatography. If the product C_pP in equation (1) is equated to G, equations (1) and (4) are identical.

The correctness of equation (4) was shown by Polson and Katz[17] in an experiment in which several proteins of known molecular weights were chromatographed singly through a column packed with 10% agarose pearls. The distribution coefficients were calculated from the elution and void volumes.

At this stage it must be stressed that the proteins used in the experiment all had low frictional ratios f/f_0 and that particles with high f/f_0 ratios will have G values which will show deviations from the values of Table IV.

Comparison of Exclusion by Polyethylene Glycol and by Gels in Gel Exclusion Chromatography

In a number of ways exclusion by PEG is similar to exclusion by gels in gel exclusion chromatography. To mention a few:

(a) If the particles (viruses or giant proteins) are large they do not enter the coils of the polymer but will remain in the intermolecular watery spaces. Similarly, if the molecules are too large for the pores, they do not diffuse into the gel particle. In both cases their distribution coefficients will be small or will approximate zero.

(b) If the substance has a very low molecular weight, e.g. monosaccharide, the substance will diffuse in and out of the polymer coils so that the distribution coefficient $\frac{C_i}{C_\epsilon}$ will be approaching unity and will not be excluded to any useful degree by the polymer. In gel exclusion the distribution coefficient will also approximate unity.

(c) If the substance is of molecular size which will enable it to diffuse into the coils of the polymer to a limited degree, its distribution or partition coefficient between the solvent and the solvent incorporated in the coils of the polymer will depend upon the degree of folding of the polymer which in turn will depend upon the concentration and chain length of the polymer molecule. Similarly, in exclusion chromatography the concentration of the gel determines the partition coefficient between the gel particles and the medium which in turn determines the position in the elution diagram.

If a gel particle, as used in gel exclusion, is reduced in size, it will at no particle size cease to partition a protein between its folded fibrils and the medium in which the gel particle is suspended. If the gel particle is still further reduced in dimension and the size of a randomly coiled PEG molecule is reached, the protein will still be able to partition itself between the gel and the medium. The only difference between the gel and the polymer particle will be that the polymer will be able to change its configuration continuously through the absence of intramolecular bonding whereas the gel particle will not, as the shape of the latter is fixed by, for example, hydrogen bonding.

Polyethylene Glycol of Molecular Mass Lower Than 6 000 Daltons

The PEG of molecular masses 6,000 D and higher seem to behave identically in displacement characteristics. Therefore it is tempting to assume that when in solution their water-filled coils are folded to the same degree. The PEG varieties of lower molecular mass are not as effective in displacement precipitation as the higher molecular varieties. It must therefore be concluded that at equal concentration (g/vol), their coiling and folding are less than those of the PEG varieties of Mr 6 000 D and higher and that water is not trapped so readily; also, because of their more open structures, the interference factor between the protein molecules and the PEG folds is smaller causing the polymer factor P to be smaller. On increasing the polymer concentration the folding will be more and their displacement characteristics will eventually be similar to those of the higher molecular types. This is demonstrated in Fig. 1 in which the precipitation indices of PEG of widely different molecular weights on gamma globulin are given.

A useful comparison may also be made with agarose gel. A gel made of 1.5% agarose of molecular mass 150 000 D left on a horizontal glass plate will retain the water held in the folds of the gel matrix, but a gel made of degraded agarose, i.e. 40 000 D of the same concentration is "sloppy" and loses its containing water rapidly when left on a horizontal glass plate. This is possibly due to less random folding and coiling of the constituent filamentous carbohydrate chains than in the case of the agarose gel made of the high molecular weight carbohydrate polymeric material. The agarose of high and low molecular weight may be compared with PEGs of long and short chains respectively.

Prediction of Displacement Precipitation

According to the present theory the following predictions may be made:

(a) The maximum rate of displacement from solution $\frac{dc}{dPEG}$ is a function of the molecular weight of the displaced substance, being zero for low molecular substances such as mono and disaccharides to infinity for very large particles or particles with large Stokes radii.
(b) In a mixture of components with different particle sizes but of equal charge densities and concentrations the component of largest particle size will be the first to be displaced from solution.
(c) The ability or concentration of PEG required to displace a substance from solution is dependent upon the molecular weight of the PEG. This in turn is proportional to the length of the polymer chain which finally determines the extent of coiling of the PEG molecule. It may be expected that the compactness or coiling of the PEG will be approximately the same for PEG of molecular weight higher than a critical value. From this it may be inferred that the displacing ability of PEG of molecular weight higher than a certain value will be identical. In the present instance the molecular mass of PEG above which no change of displacement ability with polymer molecular mass occurs is approximately 6 000 D.

DISCUSSION AND CONCLUSIONS

From work reported here and by other authors it is concluded that the mechanism whereby a protein or virus is precipitated from solution by a filamentous polymer is displacement from solution. Juckes[8] has shown by thermodynamic reasoning that the solubility S of a protein in a polymeric solution follows the "salting out" equation

$$\log S = K - \beta^{\omega}$$

where K is a constant, β is the rate of displacement with change in concentration of the polymer and ω is the polymer concentration. Jukes observed that the factor log β is a linear function of the Stokes radius of the displaced substances; being small for low molecular weight substances and large for large molecules.

Salting out is generally regarded as dehydration and precipitation with a polymer must therefore also be a process of dehydration; although the manner in which this occurs must be different.

Ogston[5] stated that the available volume a, for a colloidal particle be it protein or virus in the presence of a linear polymer equals

$$a = \exp\left[-\pi 1 \, (r_s + r_r)^2\right] \tag{5}$$

in which 1, is the total length of the polymer per unit volume and r_s, and r_r are the radii of the solute and polymer molecules respectively.

From equation (5) it may be inferred that the available volume is independent of the molecular length of the individual polymer molecules, or, stated differently, that the available volume is independent of the molecular weight of the polymer. This appears to be the case with polyethylene glycol (PEG) of molecular masses 6 000 D and higher as these appear to displace proteins and viruses to very similar degrees. The ability to displace proteins from solution by PEGs of lower molecular weights certainly differ, to an increasing degree, with decrease of the molecular weight of the PEG. This was witnessed by the results obtained in the polymer displacement experiment on gammaglobulin in which the ability to displace the protein with PEG of different molecular weight is determined. It appears that when PEG of increasing molecular weight is used, a factor becomes more prominent which is an essential property of PEG enabling it to displace proteins from solution or to dehydrate it until it is no longer soluble. A comparison of the sedimentation behaviour of methylated cellulose fractions and PEGs both of increasing molecular weights provided the answer to this. The sedimentation coefficients of the methyl-cellulose remained constant irrespective of their molecular weights. This indicates that the molecules remained thoroughly drained in watery solution and that they are, in their most distorted form, only flexible rods. The PEGs of different molecular weights behaved differently. Their sedimentation coefficients increased with increase of their molecular weights. The figures obtained ranged from 0.0 for the 1 500 D variety to 0.895 for the PEG of 20 000 D. This was regarded as being sufficient evidence for the assumption that PEGs of low molecular weights are folded only slightly but that with increase in molecular weight the folding increases. With increase of folding, water is encased in the coils. The conclusion regarding the configuration of the polymer in solution is in full accord with the Flory[13] concept.

The folded PEG molecules with their encased water may be compared with porous gel particles used in gel exclusion chromatography. Evidence for the assumption is the qualitative similarity of the displacement of proteins from the coils, based on osmotic pressure principles and the equilibrium conditions which prevail between the protein in the exterior of the coiled molecule and that in the interior and porous gel particle in contact with a protein solution. For the gel-protein-osmotic system in equilibrium an equation was derived which is similar to that for the coiled filament in contact with a protein. Whereas the polymer-coil-protein-osmotic system could not be verified, that of the gel-particle-protein arrangement by inference could. A constancy for the factor termed gel factor G was established for a number of proteins of widely differing molecular weights.

Displacement of a substance from solution is a result of two processes:

(a) Displacement from the coiled structure by virial effects, i.e. interference of the Brownian motion by the folded filaments of the polymer; and
(b) dehydration of the displaced substances to the extent that it becomes oversaturated.

REFERENCES

1. K.R. Yanamoto, B.M. Alberts, R. Benziger, L. Lawthorne and G. Treiber, Virology *40*, 734-744 (1970).
2. T.C. Laurent, Biochem. J. *89*, 253-257 (1963).
3. T.C. Laurent, Acta Chem. Scand. *17*, 2664-2668 (1963).
4. T.C. Laurent, Federation Proc. *25*, 1127 (1966).
5. A.G. Ogston, Trans. Farad. Soc. *54*, 1754-1757 (1958).
6. A.G. Ogston and C.F. Phelps, Biochem. J. *78*, 827-833 (1961).
7. E. Edmond and A.G. Ogston, Biochem. J. *109*, 569-576 (1968).
8. J.R.M. Juckes, Biochim. Biophys. Acta *229*, 535-545 (1971).
9. R. Signer and P. von Tafel, Helv. Chim. Acta *21*, 535-545 (1938).
10. A. Polson, Kolloid Zeit. *83*, 172-180 (1938).

11. A. Polson, G.M. Potgieter, J.F. Largier, G.E.F. Mears and F.J. Joubert, Biochim. Biophys. Acta *82*, 463-475 (1964).
12. A. Polson, Prep. Biochem. *4*, 435-456 (1974).
13. P.J. Flory, "Principles of Polymer Chemistry", Cornell University Press, Ithaca and London, pp. 402 (1969).
14. M.L. Huggins, J. Phys. Chem. *46*, 151-158 (1942).
15. M.L. Huggins, J. Am. Chem. Soc. *64*, 1712-1719 (1942).
16. P.J. Flory, "Principles of Polymer Chemistry", Cornell University Press, Ithaca and London, pp. 595 (1969).
17. A. Polson and W. Katz, Biochem. J. *112*, 387-388 (1969).
18. C. Tanford, "Physical Chemistry of Macromolecules", John Wiley and Sons Inc., New York and London, pp. 192-210 (1961).

3. Isolation of Viruses by Electro-Extraction of Infected Plants*

ABSTRACT

The isolation of viruses from infected plant material by a process termed electro-extraction appeared to be a convenient and simple method of obtaining viruses in a fair state of purity. The method has the advantage over the conventional methods of virus purification that the infected plant tissue is not disintegrated and that organic solvents such as chloroform and butanol are avoided. The procedure used was demonstrated on the extraction of tobacco mosaic virus (TMV) from infected tobacco and turnip yellow mosaic virus (TYMV) from chinese cabbage plants. To obtain the virus it was found advisable to freeze and thaw the plants prior to extraction.

INTRODUCTION

The process of electrophoresis in its different applications is useful as a means for the final purification of viruses[1]. The mobility of a virus relative to that of a substance of known mobility, i.e. phenol red, and expressed as an RØ value, is a property which rates of equal importance to particle size and morphology[2]. As viruses form stable suspensions, and migrate under the influence of an electrical potential gradient, it should be possible to extract viruses from infected cells if these are lysed by the virus,

*From: A. Polson, Prep. Biochem. 7:207–215 (1977).

or, if such cells are damaged through rupturing by freezing and thawing. If this separation could be accomplished, additional importance may be accrued to electrophoresis as this would enable the extraction of viruses together with their associated proteins from infected tissue leaving the insoluble material which harbored the virus, behind. The feasibility of such an application of electrophoresis was investigated on infected plants. The experiments, here termed electro-extraction, were conducted on plants that were not treated in any manner and on similar plants that received one cycle of freezing and thawing.

MATERIALS AND METHODS

The apparatus used for the electro-extraction is shown diagrammatically in Fig. (1). It consists of a perspex tube connecting via a Quickfit joint onto a vertical section of glass tubing, which in turn is bridged to a second glass tubing with a flanged extremity onto which a short piece of cellophane tubing is tied and knotted. The upper ends of the two limbs are closed with glass stoppers. The collecting chamber was made of a modified Millipore filter holder; for details see the legend of Fig. (1) in which an exploded view is seen. When assembled the functional part of the apparatus resembles the letter H. This apparatus will henceforth be referred to as the H-tube. When in operation the two limbs of the assembly were suspended in buffer contained in double cavity electrode vessels made of perspex. The electrodes were reversible silver-silver chloride covered with saturated sodium chloride.

The buffer used in the apparatus has a pH of 8.6 and its composition is as follows: 0.0175 M H_3BO_3, 0.0087 M NaOH, 0.039 M HC1, 0.018 M NaC1.

The buffer contained 0.1% sodium azide to function as a bacteriostatic agent. Ascorbic acid and sodium diethyldithiocarbaminate .$3H_2O$ were also added to the buffer to a concentration of 0.05% to act as anti-oxidants and tyrosinase inhibitors respectively. When this enzyme inhibitor was omitted from the buffer the extracted fluid turned dark brown due to polyphenol formation.

The electro-extractions were done on 30g each of tobacco plants and chinese cabbage infected with tobacco mosaic virus (TMV) and turnip yellow mosaic virus (TYMV) respectively.

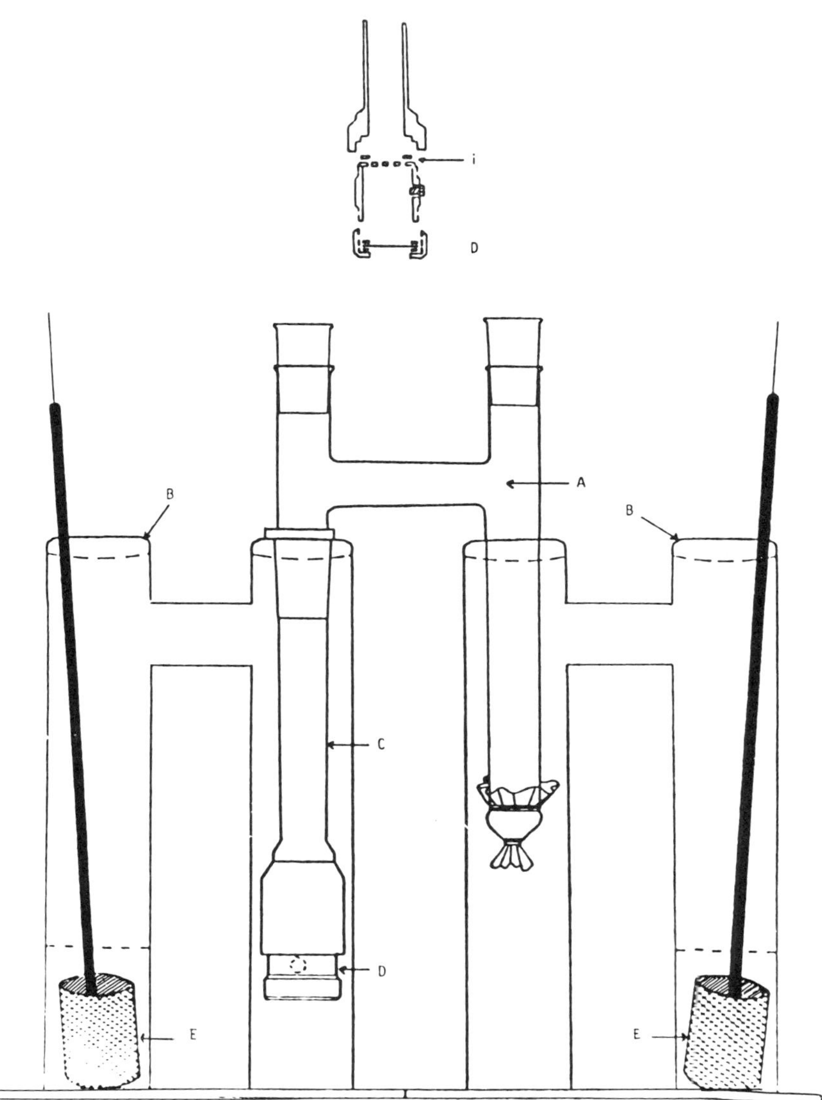

Figure 1. The apparatus for electro-extraction of virus from infected plants. The collecting chamber is shown on the right in an exploded view. Starting from below, the bounding cellophane membrane clamped between two rubber rings is shown. The ring which holds the cellophane is screwed onto the main part of the chamber. A perforated disc covers the top of the cylinder and this is screwed onto the perspex tube and is sealed off by a rubber ring. A screw plug is provided on the side allowing removal of fluid if necessary. Glass stoppers close the upper ends of the limbs of the H-tube.

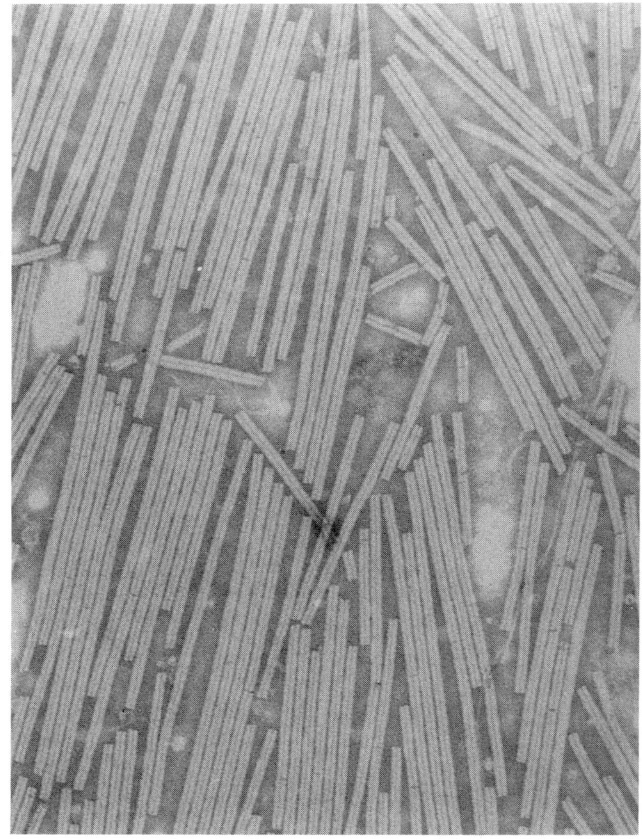

Figure 2. Tobacco mosaic virus obtained by electro-extraction from infected tobacco leaves without prior freezing and thawing. The end-on association of the virus particles was caused by the pelleting of the virus prior to electron microscopy.

OPERATIONAL

The plants were cut off just above their root levels and were washed with sterile deionized water to remove the bulk of bacteria adhering to their surfaces. They were pulled into the perspex tube. Buffer was poured into the perspex chamber to fill the collecting chamber, to replace all the air trapped by the plants and to fill the rest of the H-tube.

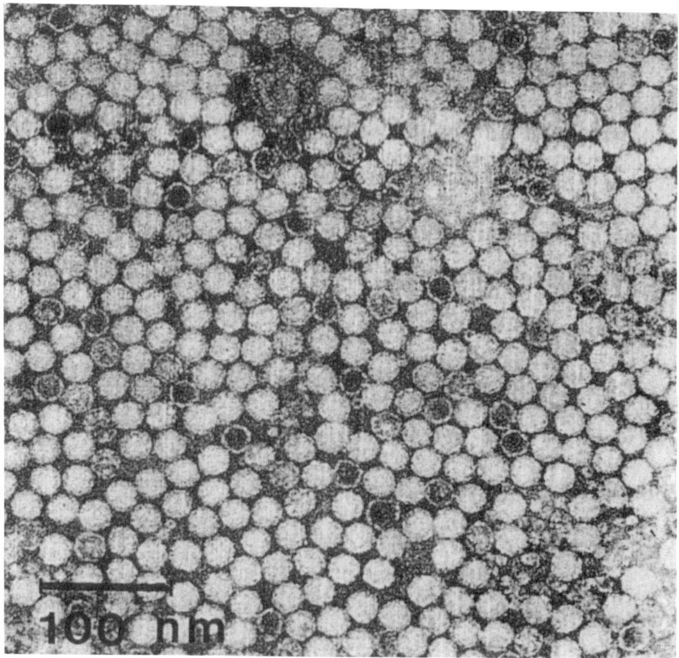

Figure 3. Turnip yellow mosaic virus obtained by electro-extraction from infected chinese cabbage leaves. The virus was obtained from the plant following freezing and thawing at temperatures -20° and 20° respectively. The virus was spun into a pellet following extraction prior to electron microscopy.

The H-tube was inserted into the vessels adjoining the electrode chambers and a current at 200 V and 25 mA passed through the H-tube allowing any liberated virus and proteins to migrate in the direction of the positive pole. As the collecting chamber is bound by a cellophane membrane, any migrating viruses, on account of their large size were stopped by the bounding membrane, while liberated chlorophyl and low molecular proteins manage to move through the pores of the semipermeable membrane.

The experiment was conducted at 4°C. After passage of the current for 24 hours the receiving chamber was unscrewed and the bounding membrane thoroughly washed with a Pasteur pipette using the buffer in the chamber to remove all the virus attached to the cellophane membrane.

RESULTS

In the experiments with TMV no difficulties were encountered with extraction of the virus from an unfrozen plant. A pellet measuring 5 mm across the centrifuge tube of the No. 40 Spinco rotor (Beckman Instruments) was obtained. The electron micrograph of the virus in the pellet indicated the presence of mainly pure virus with most of the rods associated end-on. This state of the virus may be expected when TMV is spun into a pellet, Fig. (2).

In the experiment with TYMV it was not possible to extract any virus from an unfrozen plant, but a pellet which measured 8 mm across was obtained from a single plant which was frozen at -20° and thawed prior to subjection to electro-extraction. The electron micrograph (Fig. 3) indicated that the pellet contained TYMV in a fair state of purity.

DISCUSSION AND CONCLUSIONS

A method termed electro-extraction is introduced by which viruses and other proteinaceous substances may be separated from infected plants. The method has the major advantages that the plants are not disintegrated and the extracted sap is clarified by ultracentrifugation as done when the conventional procedures are followed. In addition the customary deproteinizing method in which chloroform and butanol are used was also omitted; these solvents may have destructive effects on unstable viruses.

The isolation of plant viruses by the electro-extraction procedure may therefore be regarded as a very mild and simple method for the separation of plant viruses from their host plants.

In the case of TMV-infected plants almost pure virus may be obtained without prior freezing and thawing of the plants. It is advisable however, to include the freezing step as the yield of virus is better with little sacrifice to purity and integrity of the final product.

It is possible to concentrate virus by electro-extraction directly from the infected plant by allowing the virus to migrate onto the cellophane barrier in the collecting chamber. The concentrate may then be obtained after careful removal of the supernatant fluid.

Electro-extraction should find wide application as a means of isolating viruses that are unstable or are present in a low concentration in the plant.

ACKNOWLEDGEMENTS

The author wishes to express his indebtedness to Professor MHV Van Regenmortel and Dr. Barbara von Wechmar of the Department of Microbiology for their help in various aspects of this work.

Grateful acknowledgements are also extended to Miss Linda Stannard for the electron micrographs and Miss Gail Darroll for the illustration.

REFERENCES

1. A. Polson and B. Russel, in "Methods in Virology" vol. 2, K. Maramorosch and H. Koprowski, eds., Academic Press, New York and London, 1967, pp. 391–425.
2. M.H.V. Van Regenmortel, in "Principles and Techniques in Plant Virology", C. Kado and H.O. Agrawal, eds. Van Nostrand, Reinhold, New York, 1972, pp. 390–412.

4. Preparative Immunoabsorption Electrophoresis*

ABSTRACT

A method for separating a component from a mixture of antigens is described. The component, which may be a virus or subfraction of a virus, is isolated by driving the mixture by electrophoresis through a gel containing precipitating antibodies directed against the unwanted components. The method is illustrated by the isolation of hepatitis B antigen from whole serum and by the separation of wild cucumber mosaic virus from a strain of tobacco mosaic virus.

INTRODUCTION

Mead[1] purified the soluble antigen of rabies virus, extracted from sucking mice brains infected with the virus, by electrophoresis of partially purified soluble antigen through an antibody gradient supported in 1% agarose. The antibodies were elicited in rabbits against a normal sucking mouse brain extract. During passage of the mixture of host and viral soluble antigens through the gradient, the "normal" host antigens formed precipitin bands with their respective antibodies at optimal proportions, while the soluble antigen, to which there was no specific antibody, moved through the gel and was collected in a cellophane cup-like container at the end of the gel-filled glass tube. The separation of hepatitis B antigen (HBsAg) from serum[2] and wild cucumber mosaic virus (WCMV) from the

*From: A. Polson, M. B. von Wechmar and J. W. Moodie, Immunol. Comm. 7:91–102 (1978).

cucumber virus 4 strain of tobacco mosaic virus (CV4)[3], was accomplished in a newly designed apparatus which does not require the presence of a preformed antibody gradient.

APPARATUS

The improved apparatus is illustrated in Fig. 1. It consists of two main parts, the one being a "perspex" tube onto which a receiving chamber, bounded by a cellophane membrane at the lower end, is screwed. The other part consists of two glass tubes joined by a short horizontal tube of similar diameter. The shorter limb joins the perspex tube by way of a Quickfit joint assembly. An "exploded" view of the collecting chamber is shown separately in Fig. 1. The longer glass limb is closed by means of a cellophane tube knotted distally and tied onto the glass limb with surgical cord. The functional part of the apparatus resembles the letter H and the process of electrophoresis in this apparatus is referred to as H tube electrophoresis. The electrode vessels are two perspex cylinders joined by a horizontal tube two cm in diameter. Reversible silver-silver chloride electrodes in saturated sodium chloride solution are used in place of the conventional platinum electrodes as Pt electrodes deplete buffer salts and cause a change of pH. An additional disadvantage is the formation of "toxic" products such as chlorine gas. When silver-silver electrodes in saturated sodium chloride are used, such objectional side effects do not occur. The silver-silver chloride electrodes are interchanged at 24 hour intervals to maintain them in the reversible state. The H tube apparatus can also (without antibody-containing gel) be used for the isolation of viruses by electro-extraction of infected plants[4].

ANTISERA

(i) Antiserum to normal human serum was elicited in rabbits. The human serum was shown by electron microscopy not to contain HBsAg. (ii) Antiserum to CV4, purified by the method of von Wechmar and van Regenmortel[5] was also elicited in rabbits.

FRACTIONATION OF RABBIT ANTISERA

Serum was diluted with two volumes of phosphate buffer 0.06M at pH 8.2, and the pH adjusted to 8.6 by the addition of crystalline Na_3PO_4. Pulverized polyethylene glycol 6,000 daltons (PEG) was added with stirring to a concentration of 13% w/v. The precipitate was removed from the mix-

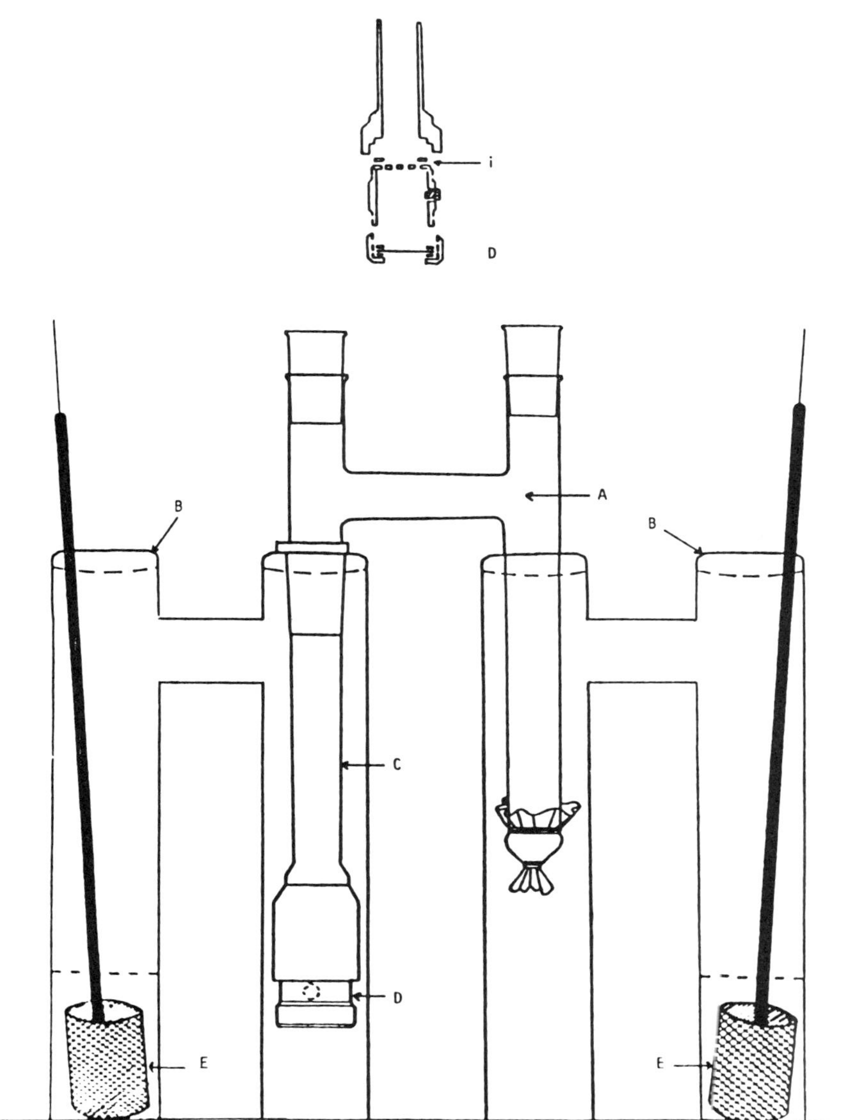

Figure 1. The H tube electrophoresis apparatus assembly showing the perspex tube and collecting chamber left of centre. An exploded view of the collecting chamber is on the right. Starting from below, the bounding membrane clamped between two rubber rings is shown. The ring screws on to the main part of the chamber. A perforated disc covers the top of the cylinder which in turn is screwed on to the perspex tube and is sealed off by a rubber ring. A screw plug is provided on the side allowing removal of fluid if necessary. Glass stoppers close the upper ends of the limbs of the H tube. Further details in text.

ture by low speed centrifugation, redissolved in one volume of phosphate buffer and reprecipitated with 13% PEG[6]. The precipitate was finally dissolved in a volume of borate buffer equivalent to one quarter of that of the serum from which it was derived. The composition of this buffer was H_3BO_3 0.035 M; NaOH 0.0175 M/ HCl 0.0075 M; NaCl 0.073 M; the pH was 8.6. The electrophoresis experiments were conducted in the same buffer.

Operational

A cellophane membrane of the appropriate diameter is placed on the perforated disc of the collecting chamber which in turn is screwed onto the perspex tube. A volume of 2 ml of partially purified immunoglobulin and an equal volume of buffer containing 2% agarose, which had been previously assessed for electroendosmosis, are mixed at 42° and allowed to congeal on the cellophane membrane on the perforated disc. This is followed by a solution of 1% agarose dissolved in the buffer at the same pH and molarity, to form a spacer gel approximately 7 cm long.

After the spacer gel has solidified the receiving chamber is unscrewed from the perspex tube, the membrane on the perforated disc removed and the chamber replaced. The gel containing the impure immunoglobulin is now in direct contact with the chamber via the perforations. The chamber is filled with the buffer through the unplugged hole on the side of the chamber, care being taken that all air is removed before the plastic screw closing the hole is replaced.

The electrode vessels and the H tube are filled with buffer and the electrodes covered with saturated sodium chloride. A potential gradient of 4 V/cm is applied across the electrodes with the polarity set in such a manner that the contaminating proteins in the immunoglobulin preparation move to the positive pole. The immunoglobulin, due to its low electrophoretic mobility is "forced" backwards by electro-endosmosis into the spacer gel in a similar manner to what happens in cross-over electrophoresis or immuno-osmophoresis.

This preliminary electrophoresis produces an antibody gradient in the spacer gel which is a consequence of the electrophoretic inhomogeneity of the immunoglobulin. After passage of the current through the gel for 24 hours, the receiving cup which at this stage contains the proteins which contaminated the immunoglobulin, is emptied and thoroughly washed, reattached to the perspex tube and filled with buffer.

The suspension of antigens containing the component to be isolated by H tube electrophoresis is mixed with an equal volume of 2% agarose and allowed to congeal on the spacer gel. The H tube is refilled with buffer and the components allowed to migrate under a potential gradient of 4 V/cm into the spacer gel containing the "gradient" of antibodies. During migration of the material through the gel those components to which there are precipitating antibodies in the spacer gel are precipitated as discrete bands while the component to be separated, to which there are no antibodies, moves into the collecting chamber. At the end of the run the chamber is unscrewed and the contents thoroughly triturated in the chamber with a pasteur pipette prior to collection. If the identity of a precipitin band is to be evaluated the gel may be pushed out of the column onto a glass surface and the agarose gel containing the band is carefully excised, frozen at -20°, thawed at 30° and compressed with a flat object. After freeze-thawing and compression of the gel approximately 80-90% of the fluid contained in the gel may be recovered. The liquid obtained in this manner contains the antigen-antibody complex. This material may be washed by centrifugation in the appropriate buffer and examined in the electron microscope. If the precipitate is that of a virus or virus capsid with its specific antibody, the complex may be easily recognized.

RESULTS

Hepatitis B Antigen

The method was applied to the isolation of HBsAg using anti whole human serum as precipitating antibody. HBsAg prepared by this method was not found to be contaminated by normal human serum components when tested against horse anti-human serum by immunoelectrophoresis. In addition, human IgG and IgM were not detected in the purified sample by radial immuno-diffusion.

Plant viruses

The partially purified mixture of WCMV and CV4 (Fig. 2) was subjected to H tube electrophoresis through gel containing a gradient of anti CV4 immunoglobulin. The duration of electrophoresis was 48 hours. During passage of the viruses through the gel an intense precipitin band was formed by the CV4 and its precipitating antibody and the WCMV, to which there were no antibodies in the gel, moved into the collecting chamber.

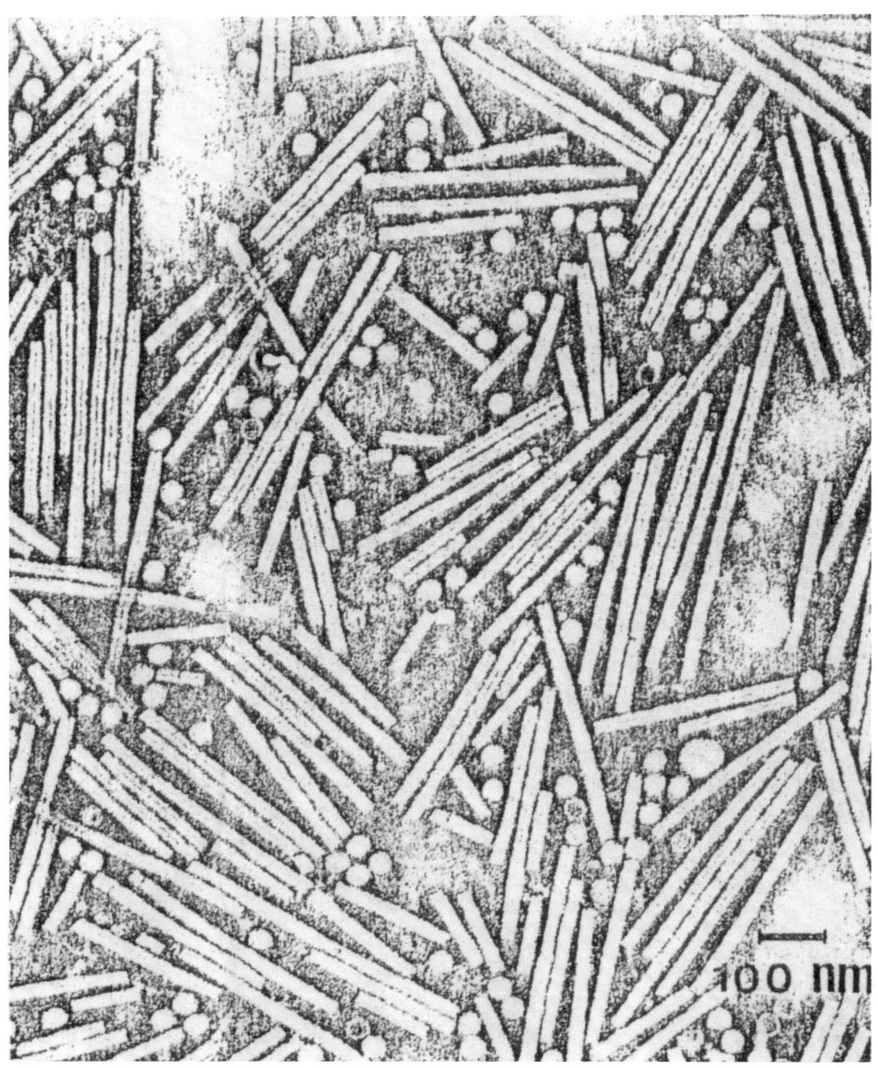

Figure 2. Electron micrograph of partially purified mixture of CV4 and WCMV prior to H tube electrophoresis.

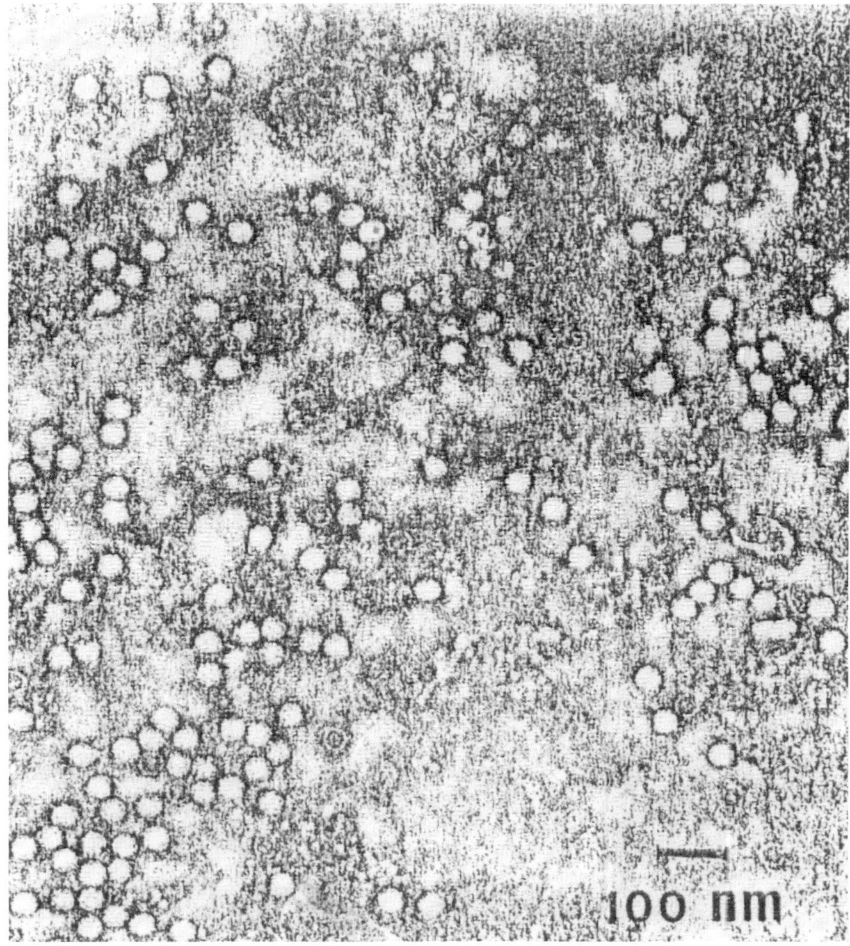

Figure 3. Electron micrograph of the WCMV isolated from the mixture by electrophoresis through agarose column containing antibodies to CV4.

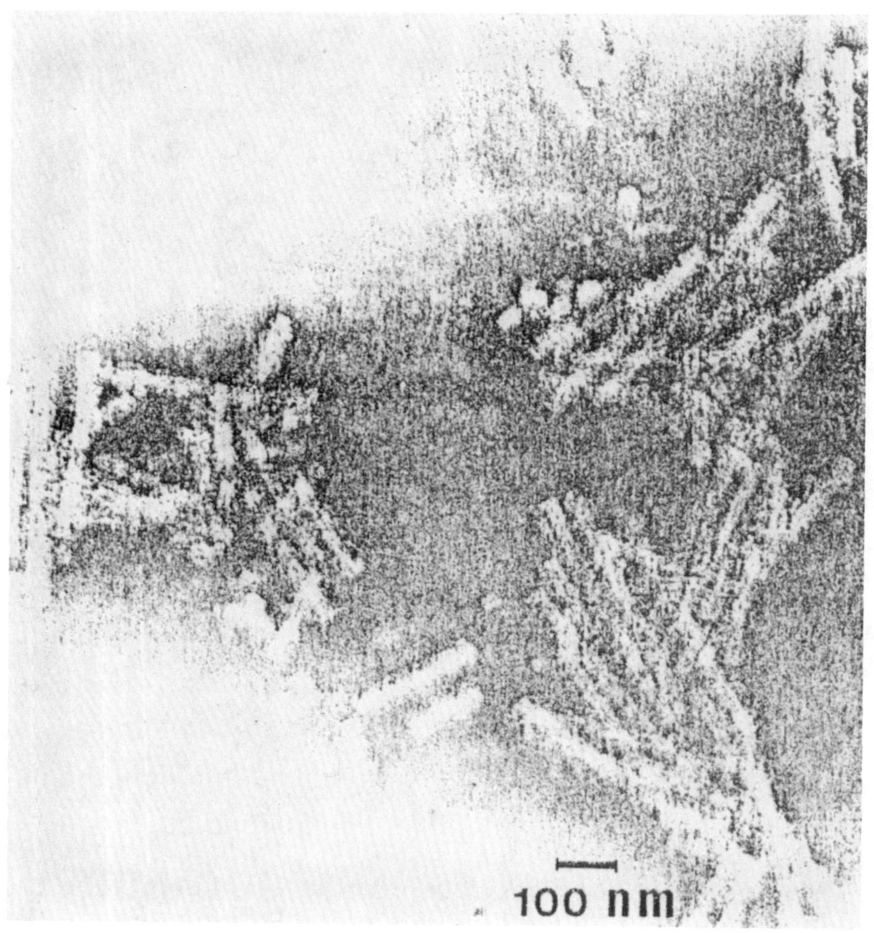

Figure 4. Electron micrograph of the immune complex of CV4 and antibodies.

After concentration of the contents of the chamber by centrifugation at 108,000 g for 90 minutes the pellet was negatively stained with phosphotungstic acid and subjected to electron microscopy. The electron micrographs show the WCMV isolated in the chamber (Fig. 3) and the immobilized immune complex of CV4 and its precipitating antibody prepared from the gel (Fig. 4).

DISCUSSION

An improved version of the preparative immuno-electrophoresis apparatus, originally used by Mead[1] for the purification of rabies soluble antigen is described. Although the conventional preparative specific absorption technique in which antibodies are attached to a solid phase may be used for the preparation of larger quantities of material, the present technique of preparative immuno-absorption has merits of its own. The main advantages are the ease in which an experiment may be conducted and the visualization of the extraneous antigens which show their presence as precipitin bands. High concentrations of HBsAg could be processed by "H tube" electrophoresis using partially purified HBsAg[7] as starting material. The HBsAg containing serum is fractionated by a PEG precipitation technique which both concentrates the HBsAg and reduces the load imposed by normal serum components on the antibody containing gel. It is important that the preliminary fractionation procedures should not expose determinant sites on the HBsAg[8] as these sites have been shown to react with anti-human serum components both by affinity chromatography[9] and by radio-immunoprecipitation techniques[10]. The concentrate produced by PEG fractionation was shown by immune electron microscopy to be non-reactive with the anti-human globulin.

The method was used as a final step in the purification of Hepatitis B antigen from serum and in the separation of wild cucumber mosaic virus from CV4 present in a plant extract.

ACKNOWLEDGEMENTS

The authors wish to thank Professor M. H. V. Van Regenmortel of the Department of Microbiology and Professor A. Kipps of the M.R.C. Virus Research Unit, University of Cape Town, for their interest in this work. Grateful acknowledgement is made to Miss L. Stannard for the electron micrographs.

REFERENCES

1. T. Mead, Biochem. J., *69*, 48 (1958).
2. J. Moodie and A. Polson (Abstract), South African Society of Pathologists, Congress: 33 (1973).
3. M. H. V. van Regenmortel and M. B. von Wechmar, Virology, *41*, 330–338 (1970).
4. A. Polson, Prep. Biochem., *7*, No. 3–4 (1977).
5. M. B. von Wechmar and M. H. V. Van Regenmortel, S. Afr. Med. J., *44*, 151 (1970).
6. A. Polson and J. R. Parker, Prep. Biochem., *3*, 31–45 (1973).
7. J. W. Moodie, L. M. Stannard and A. Kipps, J. Clin. Path., *27*, 693–697 (1974).
8. L. M. Stannard and J. W. Moodie, J. gen. Virol., *30*, 269–272 (1976).
9. A. R. Neurath, A. M. Prince and A. Lippin, Proc. Nat. Acad. Sci. (U.S.A.), *71*, 2663–2667 (1974).
10. C. J. Burrell, J. Gen. Virol., *27*, 117–126 (1975).

5. An Insert for the Reorienting Gradient Rotor for Virus Extraction from Infected Plants*

ABSTRACT

A device made of nylon and which is inserted in the bowl of the reorienting gradient rotor is described. This insert acts as a container for holding infected plant tissue for the purpose of separating virus by centrifugal force from intact and fresh plants, from frozen and thawed plants and from plant tissue disintegrated by mechanical means. The extracted fluid represents 80 to 90% of that present in the untreated plant tissue. Electron micrographs taken of concentrates of the viruses prepared by the centrifugation extraction procedure indicate that extraneous materials are reduced to a low level.

INTRODUCTION

The conventional manner for isolation and purifying viruses from infected plant material is to disintegrate the plant tissue in a Waring blender in the presence of the appropriate buffer of the preselected molarity and hydrogen ion concentration. To prevent the formation of polyphenolic compounds, anti-oxidants such as ascorbic acid and sodium thiosulphite are added. In addition tyrosinase inhibitor is also added. In general, to attain

*From: A. Polson, Prep. Biochem. 9:427–439 (479).

the best possible disruption of the plant material, the volume of fluid added is several times the volume of the plant tissue.

After disruption the homogenate is filtered through cheese cloth to remove the coarsest particles of the slurry. The filtrate is then centrifuged at a rotor velocity which is inadequate to remove the virus from the extract but which removes the bulk of the plant tissue. The supernatant fluid (SNF) is shaken with chloroform or a mixture of chloroform and butanol. This treatment causes the denaturation of the extraneous protein which becomes emulsified in the organic solvent. The emulsion is subjected to centrifugation and three layers are formed. At the bottom of the centrifuge tube a clear layer of the organic solvent, chloroform and butanol is separated. On this layer is a second which is a creamy semi-solid emulsion of organic solvent and denatured protein. Resting on this layer is the deproteinized aqueous layer containing the virus suspension.

To obtain the virus in a concentrated state it is then necessary to centrifuge it out of suspension into a pellet from which it may be redispersed into a small amount of the appropriate buffer. To reduce the work involved in the extraction and purification of a virus from infected plant material the reorienting gradient rotor bowl was fitted with an insert which proved to be helpful during the initial extraction and clarification states of the virus.

MATERIALS AND METHODS

The core of the reorienting rotor with the septa is replaced with an insert cut from a solid block of Nylon. This material was selected on account of its favorable mass to tensile ratio and because of its low heat conductivity coefficient. The insert has eight radial septa and a cylindrical inner cavity of 500 ml capacity. In the centre of the cavity is a vertical rod, one cm in diameter, having a capillary which is drilled through its centre. This narrow tube allows removal of fluid from the base of the rotor bowl with the aid of a truncated syringe needle. The wall of the cavity is perforated by eight radially directed holes of 2 mm diameter drilled through the spaces between the septa. The holes provide communication between the inner cavity and the spaces between the septa and the bowl of the rotor. The holes are in the upper half of the rotor insert.

Eight curved slabs cut from a cylinder of Nylon with their outer convex surfaces exactly corresponding to the inner surface of the rotor bowl are inserted between the septa. The concave surfaces of the slabs are machined in a sawtooth fashion which provide surfaces slanting at 45° with the horizontal. The slabs are identical and therefore interchangeable. The function of the slabs with the sawtooth ridges on the concave faces is to collect any fine debris which is centrifuged from the suspension. When removing the fluid after centrifugation extraction the fine particles remain within the sawtooth spaces.

Before the rotor is used for virus extraction the wall of the inner cavity is lined successively with coarse plastic net of mesh size approximately 3 mm adjacent to the wall, with Nylon gauze of mesh 1.0 mm and finally with a strip of nylon cloth of mesh 0.1 mm. A layer of hardened filter paper may be used in place of the 0.1 mm mesh nylon. Details of the rotor and insert are provided by the legend to figures 1 (a) and (b).

The infected plant material may be frozen at -20°C before extraction when the virus concerned is insensitive to freezing and thawing. In such cases the leaves of the plants are inserted into plastic bags and antioxidants and tyrosinase inhibitor added. Buffer is added to moisten the plant material and to dissolve the substances or virus protecting agents. After thawing the plant material is inserted in the central cavity, care being taken to have it well distributed round the central rod to minimize imbalance of the loaded rotor. It must be emphasized that the infected plants or leaves are not disintegrated by mechanical means prior to extraction. It is relied upon that the freezing and thawing process disrupts the infected cells and that the centrifugal force moves the fluid and virus particles from the damaged plant cells. After replacing the lids of the Nylon cavity and rotor bowl the assembly is ultracentrifuged at 12,000 g for 10 min in a Beckman preparative centrifuge.

When the rotor has stopped, semi-dry plant material amounting to 10-15% of the weight of the original plant material is left as residue in the inner cavity of the insert. The fluid that had been removed from the infected plant tissue by the centrifugal force is contained on the base of the rotor. This fluid is free of coarse suspension and on the serrated slabs on the surface of the bowl of the rotor a thin layer of particles that had been dislodged from the plant tissue is found. The extracted plant fluid containing the virus and protein is then concentrated by centrifugation in an appropriate rotor. The purpose of extracting frozen and thawed "whole" plants is to minimize the extraneous matter in the extracted fluid.

If the virus is damaged by the freezing and thawing procedure, the infected plant tissue is homogenized in a Waring blender in the presence of

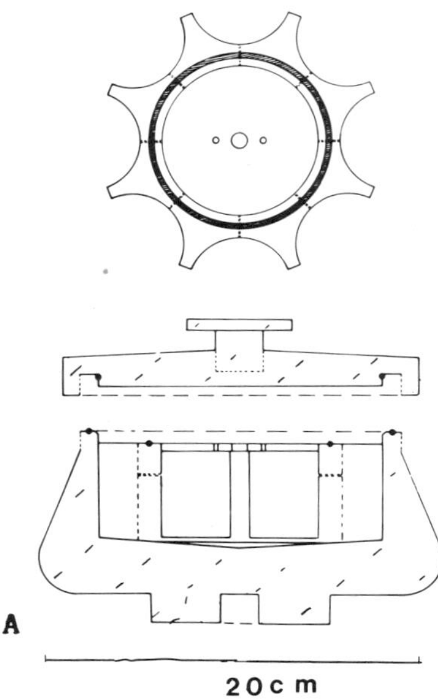

Figure 1(a). The extraction rotor; the striated part is the rotor bowl and lid which screws together. The cross-section of the rotor insert is shown in position. Note the position of the communicating holes between the bowl and the cylindrical cavity of the insert.

Upper drawing is the view of the insert from above. The 0 rings necessary for water tight sealing are shown. A hole 2 mm wide is drilled through the central rod to facilitate removal of the extracted fluid by syphoning. The hole is not shown in the diagram.

the necessary anti-oxidants and tyrosinase inhibitor. The volume of buffer added to assist the disruption of the plant tissue is kept to a minimum. The thick slurry of homogenate is poured into the cup of the insert and the assembly ultracentrifuged as described above. For this operation it is also necessary to line the wall with the layers of graded filtering materials as was done with the infected frozen and thawed plants. The SNF resulting from this operation is also free of suspended plant tissue and on the surface of the sawtooth slabs a dark green compact layer of very fine plant material is

Figure 1(b). A photograph of the rotor bowl with the insert and curved slabs; the sawtooth ridges on the inner concave surfaces are clearly seen on a slab raised from its seat for the purpose of photography.

deposited. To minimize contamination of the extracted fluid with the fine green plant debris deposited in the serrations, the fluid was syphoned off gently through a long truncated syringe needle inserted in a hole 2 Mm wide, drilled through the central rod of the insert. This hole is not shown in Figure 1. Again the plant residue on the graded filters is only slightly moist. A volume of 500 ml plant homogenate may be treated in one operation.

Some viruses, notably those belonging to the tobamo group may be extracted by centrifugation from infected plant material which has had no prior treatment besides a rinse in saline to remove the bulk of the bacteria which invariably are present on the leaves. It is fortunate that these viruses occur freely in the sap of the leaves and that by removing the sap by centrifugation the damaging effect on the virus rods caused by shearing stresses as occur when the plant tissue is disintegrated in a Waring blender, is avoided.

The virus in the extracted fluid may be concentrated into pellets by ultracentrifugation at the appropriate rotor velocity. If the volume of the extracted fluid is too large, the virus may be precipitated with polyethylene glycol MW 6,000 (PEG) and the resulting precipitate spun from suspension at low speed. The minimum concentration of PEG required would vary from one class of virus to another. In general 3-10% PEG is adequate. The precipitate is dissolved in a suitable volume of buffer and further purified by the customary pelleting using the preparative ultracentrifuge.

When dealing with filamentous viruses such as those that are classified under the potex group, pelleting and redispersion of the pellets by titration cause rupturing of the filaments resulting in loss of infectivity[1]. This undesirable effect may be eliminated by centrifuging the virus containing fluid in the thin layer centrifugation rotor[1] at 1/4 the centrifugal force and in 2/3 of the time required to pellet the virus in the conventional preparative rotors. The optimum volume of the fluid required for one operation is 120 ml and the virus is concentrated in 1-1.5 ml amounting to a hundred-fold reduction of the volume. The virus concentrate may be "washed" by dilution in the appropriate fluid and reconcentrated in a similar volume of fluid using the thin layer rotor.

RESULTS

Electron micrographs of tobacco mosaic virus (TMV) turnip yellow mosaic virus (TYMV) and cauliflower mosaic virus (CaMV) are presented in Figures 2, 3 and 4 respectively. The electron micrographs of the viruses were made on material extracted from the host plant tissue following different pretreatments of the material. TMV was obtained from infected leaves by centrifuging the intact fresh leaves. TYMV could not be extracted from fresh leaves as was done with TMV but it could be obtained from frozen and thawed leaves without further disintegration of the tissue. CaMV occurs mainly in inclusion bodies in the host plant cells and to free the virus from the inclusion body matrix it was found necessary to disintegrate the plant tissue in a Waring blender in the presence of 1 M urea and 2.5% Triton X-100[2]. Mr. D. du Plessis of the Department of Microbiology kindly placed the raw material in the form of a pulp at the authors disposal for purification by the centrifugation extraction procedure.

DISCUSSION AND CONCLUSION

An insert for the bowl of the reorientating gradient rotor is introduced. This device expedited the extraction of viruses from infected plant tissue.

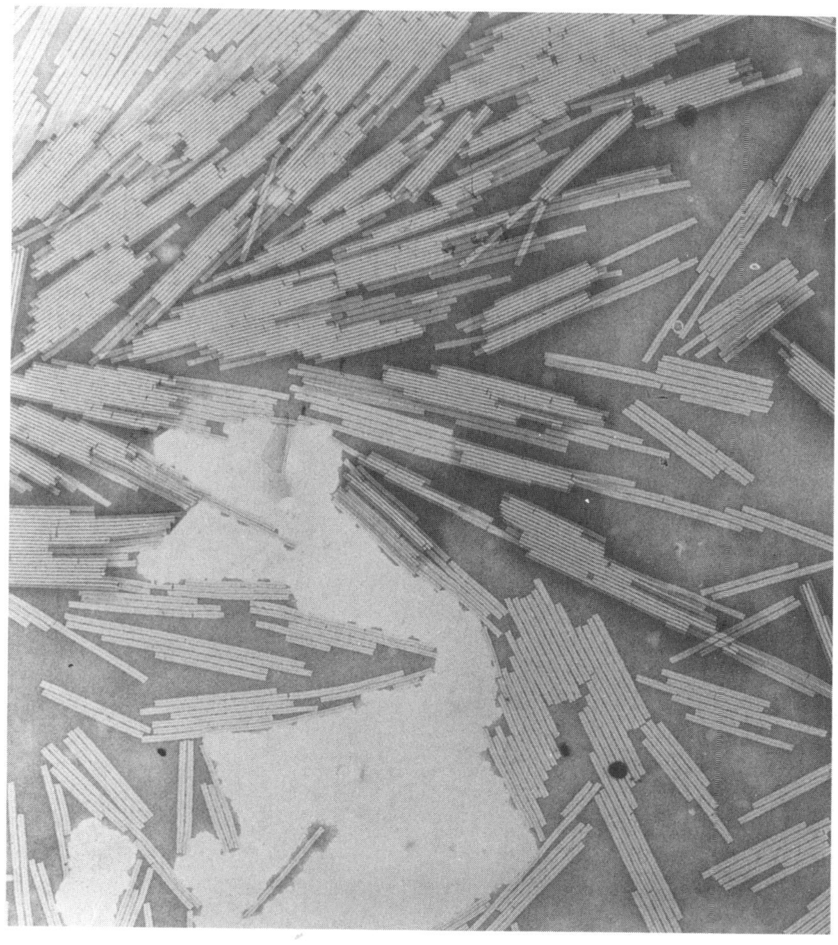

Figure 2. Electron micrograph of tobacco mosaic virus extracted by centrifugation from fresh infected tobacco leaves and concentrated by centrifugation into a pellet at 108,000 x g for 60 minutes. Negatively stained with phosphotungstic acid (PTA).

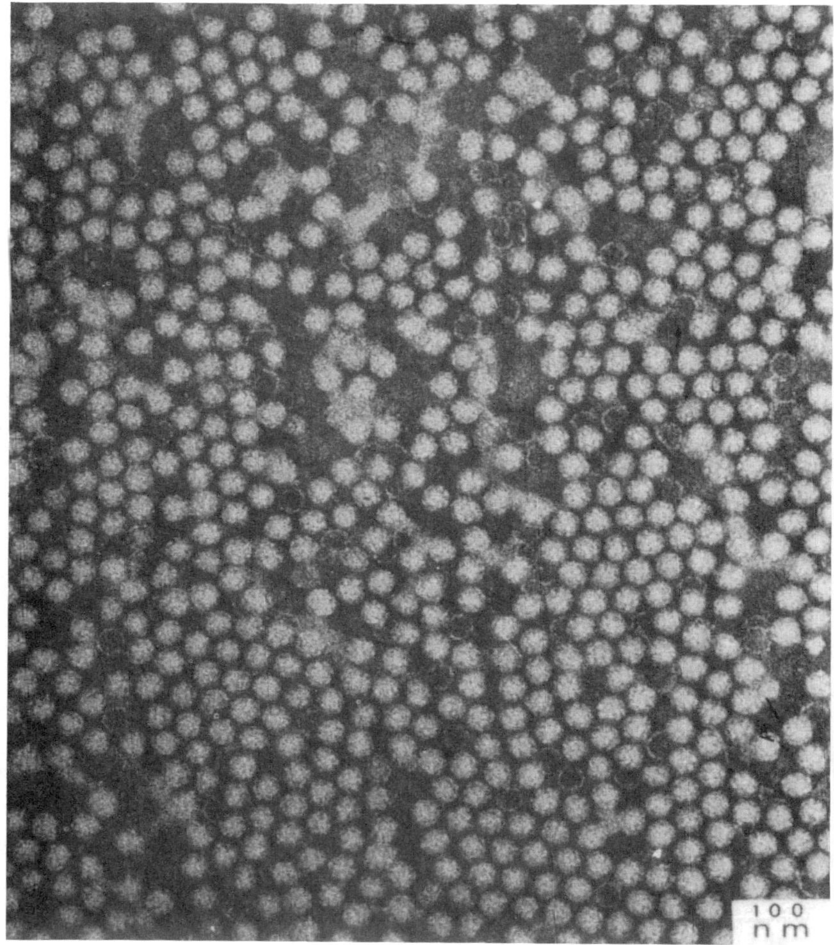

Figure 3. Electron micrograph of turnip yellow mosaic virus following extraction by centrifugation of infected leaves which received a preliminary freezing and thawing treatment. The virus was concentrated into a pellet by centrifuging at 109,000 x g for 90 minutes. Negatively stained with PTA.

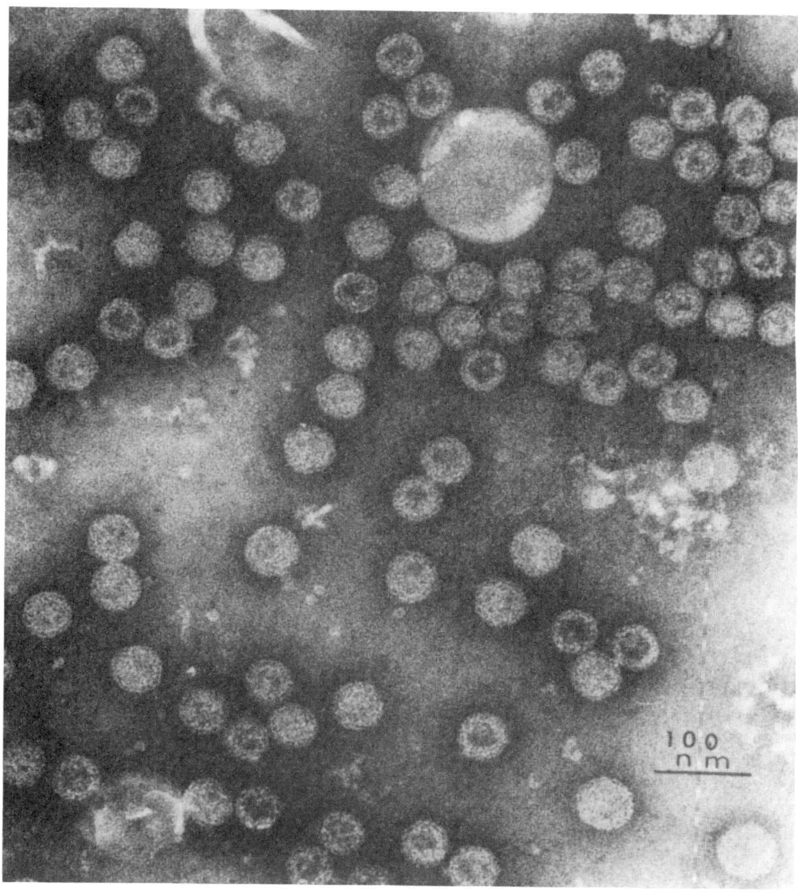

Figure 4. Electron micrograph of cauliflower mosaic virus obtained by centrifugation of infected leaves homogenized in the presence of 1M urea and 2.5% Triton X-100. The extracted fluid was spun at 12,000 x g for 10 minutes to remove some coarse particles prior to spinning into a pellet at 108,000 x g for 60 minutes. Negatively stained with PTA.

If the viruses are unaffected by freezing and thawing infected plant material may be subjected to freezing and thawing prior to extraction and the disintegration of the plant tissue in a Waring blender may be avoided. The virtues of this procedure are several but the main advantage is that very clear infective sap may be extracted and the small amount of coarse plant debris which does come free is deposited in sawtooth serrations in loose slabs on the wall of the rotor bowl.

If the virus to be extracted is sensitive to freezing and thawing the infected plants may be disintegrated in a minimum of buffer containing tyrosinase inhibitor (dithiothreitol) and anti-oxidants and the thick pulp may then be treated as was done with the intact plants. The results are somewhat different in that a thick green paste of very fine material is deposited in the serrations on the wall of the rotor bowl.

A set of filtering materials is used in the rotor insert. The open mesh filter nearest the wall of the insert serves the purpose of providing space for the extracted fluid to move in an unimpeded manner over the surface of the cylindrical cavity to the radiating holes between the septa. The second filter, being finer but still fairly rigid, prevents the spaces of the open mesh and the insert wall from being blocked and preventing fluid movement. The finest and inner-most filter retains the coarse plant residue during extraction.

The method of extraction was applied to the purification of the plant viruses TMV, TYMV and CaMV. The main virtues of the method are the very high percentage yield of plant sap which is not contaminated by plant debris and the ease with which virus may be extracted from the infected plant material.

ACKNOWLEDGEMENTS

The author wishes to express his gratitude to Professor C. von Holt and Dr. M. Barbara von Wechmar for their interest and assistance in the work, to Dr. Linda Stannard of the MRC Virus Research Unit, University of Cape Town for the electron micrographs and to Mr. A. Kiefer for the manufacturing of the rotor and insert.

REFERENCES

1. A. Polson and A. Kiefer, Prep. Biochem. *5*, (3) 199–218 (1975).
2. R. Hull, R. J. Shepherd and J. D. Harvey, J. Gen. Virol. *31*, 93–100 (1976).

5A. Concentration of Solutes by Freezing and Centrifugation

The virus extraction rotor may be usefully applied for concentrating solutes such as low molecular weight substances, proteins, virus soluble antigens and picorna viruses by centrifuging frozen dilute solutions at moderately high rotor velocities in the virus extraction rotor which had been fitted with the filters and precooled to 0°C. To accomplish this a circular stainless steel container with inner diameter equal to the inner diameter of the insert minus 5 mm was made. The reason for the smaller diameter was to allow for the thickness of the set of filters which lined the inner surface of the rotor insert. A rod attached to a flat circular metal slab closed a hole in the base of the cylinder by means of an 0 ring in the metal slab.

To conduct a concentration experiment the surface of the 0 ring was covered lightly with silicon grease and inserted in the groove on the bottom of the metal slab. The slab with the 0 ring was pressed in the position provided for on the base. The metal "cup" was filled with the solution to be concentrated and placed in a cold room operating at -20°C for a period in excess of 5 hours. At the end of this period the solution was frozen solid and had a uniform temperature across the block. The metal cup with the frozen solution was then inverted over the cavity with the metallic rod lining up with the nylon rod of the insert. By heating the outer surface of the cup with a Bunsen burner the surface of the block of frozen solution formed a thin layer of liquid and the block of frozen material slipped out of the cup into the cavity of the rotor insert. The lid of the rotor was screwed on without delay and the rotor spun at 15000 g for 10 min in a precooled Beckman No. 40 preparative centrifuge.

It is well-known that when a watery solution freezes the solute is not incorporated in the ice but is concentrated in the inter-crystalline spaces. When in such a system as the present a frozen solution is centrifuged, the concentrated solution, having the higher density moves out of the inter-crystalline regions into the spaces formed by the septa and the sawtooth slabs lining the rotor surface. When 500 ml of solution is frozen at -20°C and centrifuged in the cold rotor 90-95% of the solute is recovered in approximately 50 ml. The loss may be reduced if so wanted but this would require ultracentrifugation for a longer period which inevitably would cause some ice to thaw resulting in dilution of the solutes.

The method of concentration by freezing and centrifugation was applied to the soluble antigen of potato virus Y° (PVY°) to obtain concentrated material for zone electrophoresis in sucrose concentration gradients without resorting to precipitation with chemicals to effect its concentration as it was found that displacement of the soluble antigen with polyethylene glycol caused irreversible aggregation of the soluble antigen.

6. The Sedimenting–Droplet Procedure for Measuring Fluid Densities*

ABSTRACT

An apparatus is described which is used for the determination of velocities of falling droplets of small volumes (5 μl) through hydrophobic mixtures. From calibration curves with sodium chloride solutions of different molarities and specific gravities, the densities of the fluid in the droplets were obtained by interpolation. The density of the solutions of proteins may be determined and used in the calculation of partial specific volumes. The hydrophobic fluid used was a mixture of Dow Corning 200, kerosine to reduce the viscosity, and 1.2–dichlorobenzine to increase the specific gravity to approximately that of water.

INTRODUCTION

The technique of measuring densities of droplets of blood for the estimation of the hemoglobin or plasma concentrations by allowing droplets of the material to sink or float in salt or hydrophobic solvents has frequently been used since the pioneering work of Linderström–Lang. The literature on the subject has been reviewed by Rathjen et al.[1]. In addition these authors also introduced a refinement of the technique of the falling–droplet

*From: A. Polson, H. Alk and K. Achtleitner, Analyt. Biochem. *107*:25-31 (1980), with permission.

velocity procedure which, when applied to the determination of hemoglobin content of blood, appeared to be a reliable alternative to the conventional methods of hemoglobin assay. In this procedure the sample (20 μl) from an Eppendorf micrometer pipet was allowed to sediment through Dow Corning 200, a dimethylsiloxane polymer of density 0.934 g/cm^3 at 23 °C, in a glass tube held in a thermostated block. On passing down the tube the droplet interrupted a beam of light directed on a photoconductor. This event started a timing mechanism. Further down the tube the droplet interrupted a second beam of light which in turn stopped the timing device. The time taken by the droplet to travel from the first to the second beam was provided as a digital readout. The same linear relationship was obtained between the density of the droplet and the reciprocal of the time taken by the droplet to move from the first to the second beam irrespective of whether the solution tested was salt, plasma, or whole blood. From this it may be concluded that the dimethylsiloxane polymer did not interact with the blood plasma. However, Rathjen et al.[1] gave no details of the electronic circuit used in their apparatus.

The falling-drop method of Rathjen et al.[1] can be used only for concentration and/or density measurements when the concentration of the substance under examination is high, the main reason for this being the low specific gravity of the dimethylsiloxane through which the droplet of protein has to sediment. This results in a decrease of the sensitivity. The smaller the density differential is between the hydrophobic medium and the sedimenting droplet the greater the sensitivity and accuracy become for concentration and density determination. If a fluid could be found which is strongly hydrophobic and the density be changed by the addition of another hydrophobic substance in such a manner that the density equals that of water, very small differences in specific gravities between the hydrophobic mixture and the watery solution may be measured. These may then be correlated with the concentration of substance in the solutions.

MATERIALS AND METHODS

The hydrophobic mixture was made of dimethylxyloxane, Dow Corning 200, kerosine, and 1.2-dichlorobenzine. Kerosine was added to decrease the viscosity and 1.2-dichlorobenzine to increase the specific gravity to that of water at the temperature of operation. Lyophilized bovine serum albumin (BSA)1[1] was donated by Miles Laboratories, Cape Town. It was dissolved in 0.1 M NaCl and dialyzed as a concentrated solution, approximately 100 mg/ml, against the same medium. The concentration was determined by the

Lowry method and by light refraction methods. The falling-droplet apparatus was constructed in our laboratory and full details are given in the text with reference to Fig. 1.

Drop detector circuit [2,3]: The circuit shown was used to start/stop a Casio computer quartz CQ–1 digital stopwatch. The digital stopwatch provided the readout with 0.1-s resolution for the time taken by the droplet to traverse between two phototransistors. The Darlington phototransistors Type PH10 were mounted approximately 150 mm apart. The uppermost phototransistor was positioned 70 mm away from the droplet releasing position to allow the rate of sinkage to stabilize before it interrupted the first light beam. The glass tube T shown in the drawing was mounted in a light-proof hollow brass jacket in order to allow temperature-controlled water to circulate. An isothermal heat sink is necessary for the glass tube filled with the hydrophobic mixture to ensure that the density and viscosity of the liquid are not affected by varying temperatures.

The circuit consists essentially of two phototransistors and a comparator circuit. The dc voltage produced by the phototransistors is compared with a manually adjustable reference voltage of opposite polarity. The output of the comparator operates a high-speed Reed relay which is used to start-stop a digital stopwatch. The Reed relay contacts were wired in parallel with the start/stop push-button switch of the stopwatch. In the quiescent condition the voltages produced by the phototransistors at points A and B are offset by the manually adjustable reference voltage V_{ref} which is of opposite polarity. To prepare the circuit for operation RV_3 is slowly rotated until the comparator just operates the Reed relay. The potentiometer is then turned back fractionally in order to be just below the trigger threshold. When a droplet interrupts the light beam between the light-emitting diode (extra bright type - 40-Mcd output) and the phototransistor, the level of light reaching the phototransistor is reduced due to the scattering effect of the droplet. The current through the phototransistor consequently decreases, thereby causing the dc voltages at points A or B to increase. A voltage pulse of typically 0.5-V peak-to-peak amplitude is generated, the width of the pulse depending on the speed of the droplet, but a 2-s pulse width or longer is normally generated. The output transition of the comparator occurs when

$$\frac{V_A}{R5} + \frac{V_B}{R6} = \frac{V_{ref}}{R8}$$

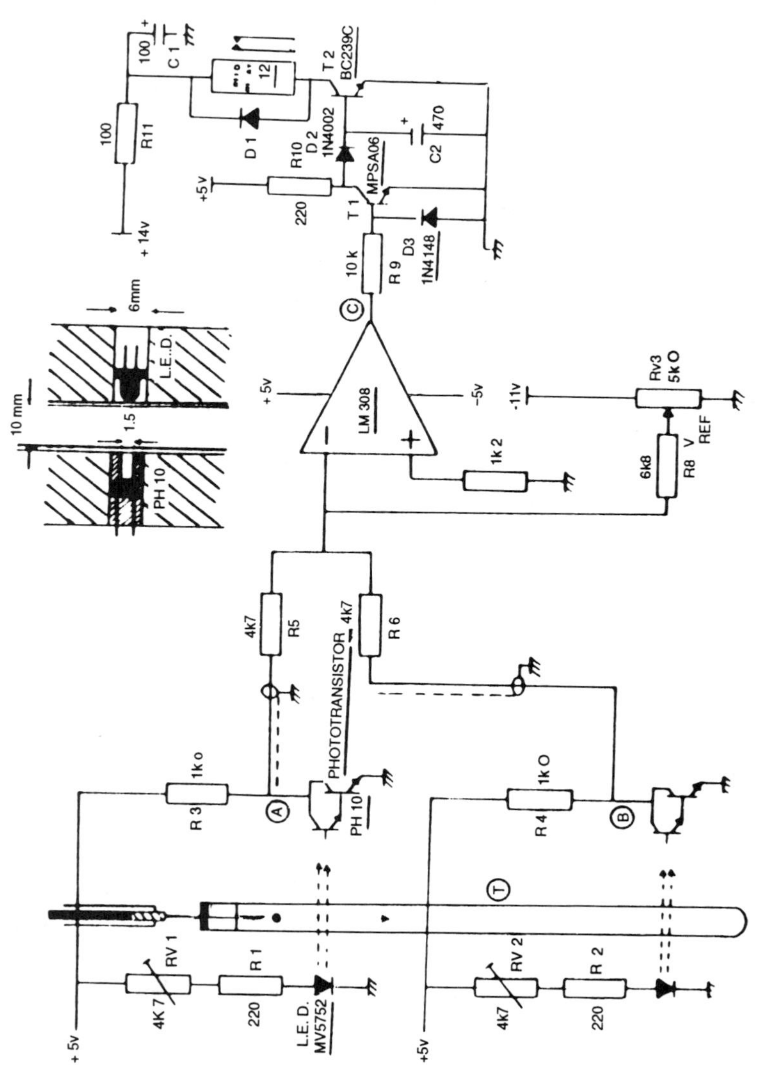

Figure 1. Drop Circuit detector. Details in text.

R8 includes the effective resistance of RV3. The threshold sensitivity is about 20 mV in the most sensitive position.

The light falling on the phototransistors via the glass tube filled with a hydrophobic mixture is adjusted by the present potentiometers RV1 and RV2 so that approximately 2.5 V is produced at points A and B in the quiescent conditions. This allows maximum excursion above and below the normal standing dc voltage.

In the quiescent condition the output voltage of the comparator at point C is about +4 V. This causes transistor T1 to be fully on and its collector emitter voltage to be only about 100 mV. Transistor T2 will therefore be off, which in turn means that the relay is inoperative.

When a pulse is generated at A or B, the output voltage of the comparator point C goes from +4 to -3 V thus turning off T1. Capacitor C2 charges rapidly via R10 and D2 thus causing T2 to turn on when C2 is sufficiently charged. Capacitor C2 now discharges via the base emitter junction of T2. Transistor T2 turns off after approximately 4s thus releasing the Reed relay. Resistor R9 is used to limit the base current of T1 and to maximize the open-loop gain of the operational amplifier. Diode D3 is used for reverse polarity protection of T1.

The leads from the phototransistors to the comparator should be coaxial cables to minimize interference.

RESULTS

The drop-sedimentation rate technique was designed and developed for measuring densities of small volumes of aqueous solutions of solutes which are insoluble in the hydrophobic mixture through which the droplets of the solutions sediment. As tables are available for the densities of sodium chloride of various molarities, these data were used for establishing a sedimentation velocity-density curve. Such a curve is shown in Fig. 2. In Fig. 3 is shown a droplet-sedimentation curve obtained by measuring the velocity of the sedimentation of droplets (5 μl) of BSA dissolved in 0.1 M NaCl and diluted serially twofold in 0.1 M NaCl and a density reference line of NaCl. The concentrations of NaCl and BSA are indicated on their respective abscissae.

Reproducibility

The data for obtaining Fig. 3 may be used to illustrate the reproducibility of the sedimentation rates (SRs) of the droplets of the

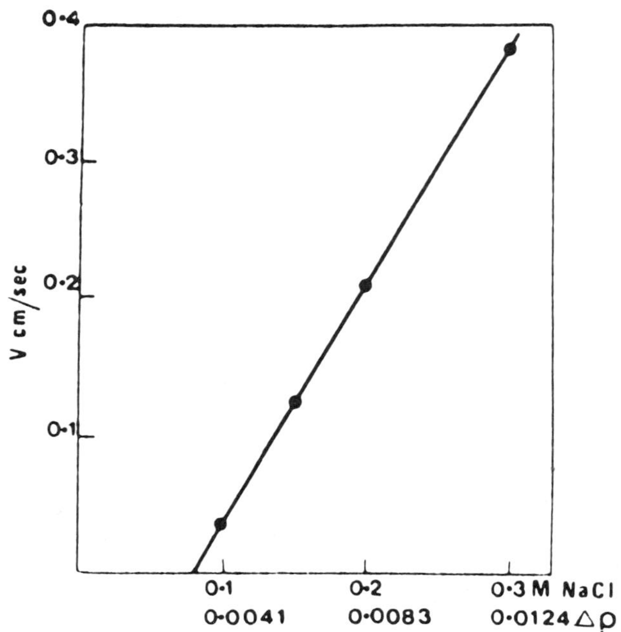

Figure 2. Sedimentation rate-density differential ($\Delta\rho$) relationship of NaCl at different molarities. Droplets (5 μl) sedimented through a hydrophobic mixture of Dow Corning 200, kerosine, and 1.2-dichlorobenzine. If required, the density of the mixture may be obtained by extrapolation to zero sedimentation rate.

different NaCl solutions. In Table 1 it will be evident that the variation of velocities of sedimentation of droplets of the various NaCl solutions occurred in the fourth decimal place with repeated measurements.

The reproducibility of the SRs of droplets (5 μl) of BSA through the hydrophobic mixture is shown in Table 2. The data for obtaining the BSA droplet-sedimentation curve in Fig. 3 are used.

Partial Specific Volume

The partial specific volume of a substance may be calculated from the equation

$$\overline{V} = 1 - \frac{(M_{su} - M_{sv})}{C}$$

in which M_{su} and M_{sv} are the masses of 100 ml of the solution and solvent, respectively, and C the concentration in grams per 100 ml. Using the data presented in Tables 1 and 2 the partial specific volume ($\overline{V}$) of BSA was calculated. This was done by interpolating the SR values of the droplets of the BSA solutions of different concentrations into the density-SR curve of

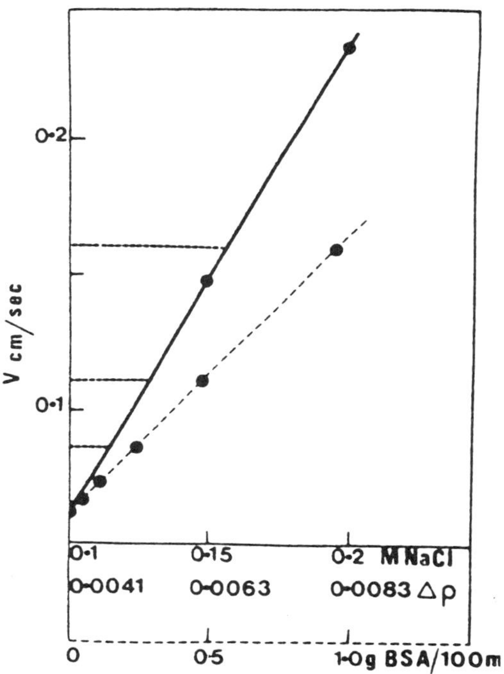

Figure 3. Inclined interrupted line, sedimentation rate-($\Delta\rho$) relationship for droplets of BSA solutions of different concentrations; solvent 0.1 M NaCl. Solid inclined line, sedimentation rate of drops of NaCl solution at different molarities; the line acts as reference density for the BSA solutions. Density differentials of BSA solutions from 0.1 M NaCl (solvent) indicated by horizontal interrupted lines. Droplet volume 5 μl. Medium through which droplets sedimented is similar to that used in Fig. 2.

Table 1. Sedimentation Rates of Droplets (5 μl) of NaCl Solutions of Different Molarities Through the Hydrophobic Mixture of Density 0.99921 G/CM^3 at 22°C

Molarity	ρ	1[a]	2	3	4	Average
0.1	0.0042	0.0606	0.0605	0.0604	0.0606	0.0605
0.15	0.0063	0.1475	0.1473	0.1475	0.1472	0.1474
0.2	0.0083	0.2343	0.2343	0.2351	0.2348	0.2346

Note. Density of 0.1 M NaCl = 1.00186 at 22°C; $\Delta\rho$, difference of density of NaCl at various molarities above that of water. The fifth decimal place is approximate as chemical tables only quote to fourth.

Table 2. Sedimentation Rates of Droplets (5 μl) of BSA of Different Concentration in 0.1 M NaCl Through the Hydrophobic Mixture

BSA concentration (mg/ml)	1[a]	2	3	4	Average
9.50	0.1587	0.1585	0.1589	0.1587	0.1587
4.75	0.1105	0.1100	0.1105	0.1101	0.1102
2.37	0.0863	0.0862	0.0863	-	0.0863
1.185	0.0742	0.0738	0.0741	0.0741	0.0740
0.592	0.0674	0.0677	0.0676	0.0675	0.0675
0.000 (0.1 M NaCl)		See Table 1			

Note. Further details in Note to Table 1.

Table 3. Partial Specific Volume of BSA from Data Presented in Tables 1 and 2

BSA Concentration (mg/ml)	$\bar{V}$ (cm^3/g)
9.50	0.746
4.75	0.744
2.37	0.734
1.185	0.735
0.592	0.732
Average	0.738

Table 4. Sedimentation Rates of BSA of Low Concentration in 0.1 M NaCl

BSA Concentration (mg/ml)	S ($\overline{V}$ cm/s) (average)
1.190	0.0510
0.595	0.0425
0.297	0.0388
0.149	0.0367
0.074	0.0358
0.000 (0.1 M NaCl)	0.0344

Note. The density of the 0.1 M NaCl was 1.00186 g/cm^3 and that of the hydrophobic mixture 1.00098 g/cm^3. Temperature 22°C. The 5th decimal in the density is approximate. Volume droplet 5 μl.

sodium chloride to obtain the density of the BSA solutions. The relationship between concentration and/or density of the NaCl solutions and the SRs is linear over the concentration range in which the measurements were made (see also Fig. 2). It must be stressed that the SR measurements of the salt and of the BSA solutions were made in the identical hydrophobic mixture.

The $\overline{V}$ values calculated for BSA at different concentrations are presented in Table 3.

Enhancement of Sensitivity

The sensitivity of the droplet-sedimentation technique was improved to allow the quantitative determination of protein concentration of the order of 0.1 mg/ml and less. This was done by decreasing the density differential between the protein solvent (0.1 M NaCl) and the hydrophobic mixture by the addition of small volumes of 1.2-dichlorobenzine, the high-density component, to the mixture. The difference in density between 0.1 M NaCl and the hydrophobic mixture was decreased to the level at which a droplet (5 μl) of the 0.1 M NaCl solution required more than 7 min to sediment through the required 15 cm of the hydrophobic mixture in the tube.

In Table 4 the SRs of droplets (5 μl) of BSA at low concentration in 0.1 M NaCl are presented.

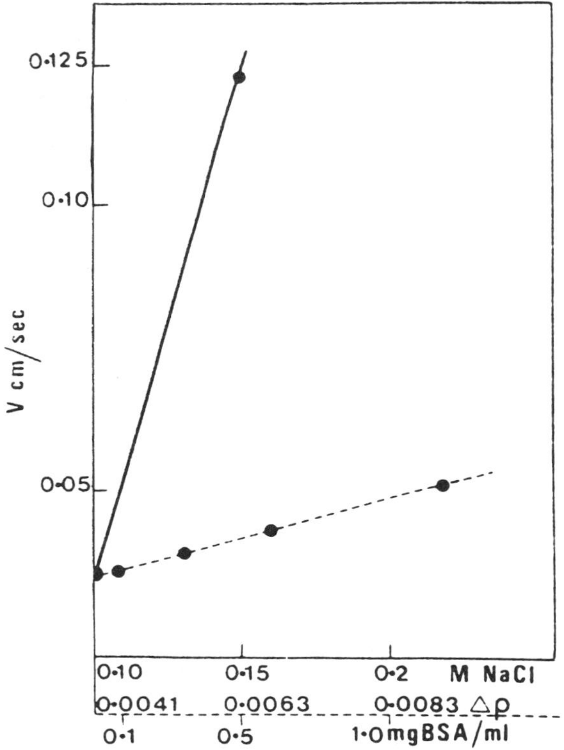

Figure 4. Sedimentation rates of droplets (5 μl) of BSA at low concentration, solvent 0.1 M NaCl, interrupted inclined line. Same hydrophobic mixture used for obtaining Fig. 3. Solid inclined line. NaCl density reference.

The data in Table 4 are also presented graphically in Figure 4 to indicate the linearity between the concentration of BSA in the lower concentration range and SRs. For comparison a line relating the sedimentation rate of NaCl at two molarities determined with the same hydrophobic mixture is presented in the diagram. The sensitivity of the technique may still be further improved by decreasing the density differential between the 0.1 M NaCl and the hydrophobic mixture, but this would be impractical as the time required for a series of measurements would then be excessively long.

It appeared that no water in the sedimenting droplets dissolved in the hydrophobic mixture or, if the water did dissolve, that the mixture soon became saturated with water drawn from the sedimenting droplets and from the watery solution which accumulated in the conical base of the tube containing the mixture. Proof for this statement can only be given by inference as the initial SRs of the droplets through a newly prepared mixture were the same after the mixture had been in contact with the watery solution in its conical base for several days.

It is not possible with the present apparatus to determine whether there is a difference in the SR of a droplet in the upper and lower regions of the hydrophobic mixture as the SRs were measured between two fixed positions which were 15 cm apart.

DISCUSSIONS AND CONCLUSIONS

The sedimentation rates of droplets (5 μl) of NaCl of different molarities and/or densities (obtainable from chemical tables) through hydrophobic mixtures has been investigated. In these experiments the density of the hydrophobic mixture was adjusted to lie between that of water and 0.1 M NaCl. The SRs were found to be a linear function of the density of the droplets and the experimental data obtained were used for obtaining a reference line from which the densities of the solutions of proteins, which were dissolved in 0.1 M NaCl, were obtained by interpolation.

The partial specific volume, $\overline{V}$, of bovine serum albumin was calculated from the density of solutions of the protein serially twofold in 0.1 M NaCl. From the values obtained it would appear that at 2.37 mg/ml and lower concentrations the partial specific volumes by the present method are in closer agreement with the accepted literature value of 0.734 cm^3/g[4]. At higher protein concentrations, values for $\overline{V}$ were obtained which were larger than Hunter's values. No explanation for the apparent dependency of $\overline{V}$ on the concentration of the BSA is offered at this stage.

As no special effort was made to determine the protein concentration used in Table 4 very accurately and, as in the difference in density between the hydrophobic mixture and the protein solutions became very small, the values for $\overline{V}$ calculated from the SR figures in this table were considered meaningless. The partial specific volume of BSA was calculated to the third decimal which may not be justified for the individual values but it was considered important to know as it would influence the average $\overline{V}$ calculated for the BSA at different concentrations and also to indicate the degree of reliability of the individual values.

The sensitivity of the droplet-sedimentation procedure could be improved to the extent where 0.1 mg/ml BSA may be estimated with confidence. This was done by decreasing the density between that of the hydrophobic mixture and 0.1 M NaCl (the protein solvent). Greater sensitivity is possible by decreasing the density differential even more, but this was not done as the time involved in obtaining an extended series of measurements would be excessively long. However by increasing the droplet size the SR may be increased.

As it is possible to measure concentrations of proteins and other solutes, which are insoluble in the hydrophobic mixture, in very small volumes, several interesting applications of the droplet-SR technique are possible. Thus the mass of viruses in pellets, obtained by ultracentrifugation, may be determined after redispersion of the pellets in media of known density; the number of immune Fab_1 pieces which may possibly bind to an antigen may be established; the elution diagrams from chromatography columns, obtained by uv absorption monitoring devices, may be converted into their protein concentration equivalents and the concentration ratios between the material which diffused across a boundary, formed between a solvent and solution, and that of the starting material, in analytical diffusion experiments, may be established.

In conclusion it may be stated that, if necessary, the solutions in which the measurements were made may be recovered quantitatively by removing the same from the conical end of the tube in which the droplets accumulated. This may be done with a narrow bore plastic tube attached to a Pasteur pipet. The salvaged material may then be used in other studies.

ACKNOWLEDGMENT

The authors would like to express their gratitude to Professor C. von Holt for his continuous interest in this work.

REFERENCES

1. Rathjen, C.P., Christner, J.E., and Baguslaski, R.C. Anal. Chem. *47*, 1863-1865 (1955).
2. Clayton, G.B. Operational Amplifiers, (1971) Butterworths, London.
3. Linear Integrated Circuits National Semiconductor Co. (1976).
4. Hunter, M. J. Phys. Chem. *71*, 3717 (1967).

7. Determination of the Number of Fab_1 Pieces Which React With High Molecular Weight Antigens*

ABSTRACT

The sedimenting-droplet technique was used for the determination of the masses of immune Fab_1 pieces which may bind to antigenic sites on three haemocyanins. Similar experiments were conducted on three viruses which belong to the Tobamo group. It was calculated that 384 homologous Fab_1 pieces could bind to a molecule of the haemocyanin of *Burnupena cincta* and 950 homologous Fab_1 pieces to a virion of tobacco mosaic virus. By taking special precautions the number of undergraded antibody molecules which can bind to a Tobacco mosaic virus particle may be determined with the same technique.

INTRODUCTION

In a previous communication[1] a method for determination of densities of aqueous solutions was described. Briefly, the method is based on the rates of sedimentation of accurately measured droplets, 0.005 ml, of the solutions through hydrophobic media with densities slightly lower than that of the dispersion media of the solutes. The reliability of the technique was demonstrated in the determination of the partial specific volume of bovine

---

*From: A. Polson and K. J. van der Merwe, Immunol. Comm. *11*:261-273 (1982).

serum albumin (BSA). The value obtained agreed well with the value on which there is general concensus[1]. Conversely, assuming the average value of the partial specific volume of proteins to be 0.738 g/ml, the concentration of protein in a medium of known density may be determined. Because of the ease with which the sedimenting droplet technique may be applied in protein and virus research, attempts were made to establish the mass of Fab_1 pieces of antibodies which could bind to the surface of large antigens such as the giant haemocyanins and some plant viruses of the Tobamo group.

MATERIALS AND METHODS

Haemocyanins

Haemocyanins were obtained from the molluscs, *Burnupena clincta*, *Fusus varicatus*, and *Haliotis midea*.

Standard BSA Solution

A solution of BSA containing 95 mg/ml in 0.1M NaC1 was used as reference standard as well as twofold dilutions of the protein ranging from 9.5 to 0.148 mg/ml in 0.1M NaC1.

Hydrophobic Medium

The medium in which the droplets were sedimented was a mixture of equal volumes of kerosene and Dow Corning 200 Fluid 20 C.S. silicone oil, to which was added 1,2-dichlorobenzene, until the density of the mixture was equal to or slightly lower than that of 0.1M NaCl, which is 1.0017 g/cc at 23°C.

This density is attained by adding 1,2–dichlorobenzene to the hydrophobic mixture until a droplet of 0.1M NaC1 remains suspended in the hydrophobic fluid after introduction. To clarify this, a brief description of the experimental procedure and the behavior of the droplet follows.

A small amount of 0.1M NaC1 is sucked into a syringe fitted with a truncated needle and the needle lowered into the hydrophobic mixture. A droplet of the salt solution is then let out while the orifice of the needle is still well below the meniscus. The droplet adheres to the needle while below the meniscus but is detached by the meniscus when the needle is withdrawn from the fluid. The droplet sinks rapidly initially but is slowed down; beyond 15 to 20 mm from the meniscus its sedimentation rate to the base of the container will be constant if the density of the droplet is higher than that of the surrounding medium. If the density of the 0.1M NaCl solution is lower than that of the hydrophobic medium the droplet will likewise sink when dislodged by the meniscus of the hydrophobic medium and will sink at a decreasing rate until a distance of approximately 15 mm from the meniscus is reached. Here the droplet will stop momentarily and immediately reverse its motion to stop directly below the meniscus. If the density of the droplet is the same as that of the hydrophobic mixture the droplet will sink to approximately 20 mm from the meniscus at which level it will remain suspended.

A hydrophobic mixture of density slightly lower than that of the droplet is desirable. This is attained when the hydrophobic mixture is poured into the observation tube; the reason being that a small amount of the component with the highest partial vapor pressure in the mixture, 1,2-dichlorobenzene, evaporates during the transfer from the mixing vessel, leaving a medium of density slightly lower than that of the droplet. A desirable rate of sedimentation of a droplet of volume 0.005 ml would be 15 cm in 150 sec at 23 °C.

The Apparatus

The apparatus used was described previously[1]. Essentially, the rate of sedimentation of a droplet (0.005 ml), delivered by a Hamilton syringe with the aid of a micrometer screw, is measured over a distance of 150 mm in a cylindrical glass tube of 10 mm internal diameter. The temperature of the glass tube and its contents was kept constant at a level approximately 1 °C above ambient by an insulated metal jacket, through which water was circulated from a thermostatically controlled reservoir. To ensure that the temperature was not influenced by a standard pumping device, an air lift was used for circulating the water. The contents of the reservoir were stirred with a magnet on the base of the reservoir.

Viruses

Three samples of viruses belonging to the Tobamo group were kindly supplied by Dr. D. du Plessis of the Department of Microbiology, University of Stellenbosch. The viruses were Yellow Tomato Atypical Mosaic Virus, (Y-TAMV), Green Tomato Atypical Mosaic Virus, (G-TAMV)[2] and the common strain of Tobacco Mosaic Virus (TMV).

Antisera

Antiserum to TMV was elicited in a rabbit and that against the haemocyanin of *B.cincta* was isolated and purified from the yolks of eggs from hens which were immunized with the haemocyanin[3]. The rabbit immunoglobulin (anti TMV) was partially purified by precipitation with polyethylene glycol MV 6000 (peg). It is understood that both types of specific antibodies represent unknown percentages of the total immunoglobulin IgY and IgG isolated.

Enzyme

Mercury papain (Worthington) was a gift from Clinical Science, Medical School, Cape Town.

Fab_1

Fab_1 was prepared by the technique described by Porter[4]. Hen Fab_1 anti *B.cincta* haemocyanin contained 8.2 mg/ml and rabbit anti TMV, Fab_1 9.3 mg/ml.

Immuno-Electron Microscopy

Electron micrographs of the free antigens and antigens complexed with Fab_1 (immuno-electron microscopy) were kindly made by Mrs. M. Hodgkiss of the Virus Research Unit, Medical School, Cape Town.

Analytical Ultracentrifugation

The hen immunoglobulin and the Fab_1 prepared from it were subjected to analytical ultracentrifugation in the Beckman Model E using the plain and wedge window cells respectively. The rotor velocity was 56,000 rpm and exposures were taken at 32 min. intervals. The ultracentrifugation was done at 20°C.

Determination of Concentration of Fab_1

Because of several uncertainties involved in the estimation of concentration of Fab_1 by ultra violet light absorption at 280 nm, determination of concentration by analytical ultracentrifugation was considered more reliable. Briefly, BSA, containing 9.5 mg/ml as reference standard and Fab_1 solution were ultracentrifuged concurrently in the plain and wedge window cells of the AnD rotor respectively. Their sedimenting boundaries were recorded photographically using the Schlieren optical system (Fig. 2). By enlarging and tracing of the Schlieren diagrams obtained after 48 min. centrifugation at 56,000 rpm, the respective areas included by the base lines and the Gaussian distribution-like sedimenting diagrams were determined. Using the area obtained for the BSA at 9.5 mg/ml as standard, the concentration of the Fab_1 was calculated. The error peculiar to this method would be dependent upon the ratio of the sedimentation coefficients of the standard substance to that of the component of unknown concentration; this error may be eliminated. As BSA has a sedimention coefficient of 4.1 at 9.5 mg/ml and the Fab_1, 3.7 Svedberg units, the concentration of Fab_1 was calculated from the equation:

$$\frac{\text{Area (BSA)} \quad 3.7}{\text{Area } Fab_1 \quad 4.1} \cdot 9.5 = Fab_1 \quad \text{mg/ml}$$

It must be stated again that the specific immune Fab_1 is a fraction of the total.

Determination of the Number of Undegraded IgG Molecules That Can Bind to a TMV Virion

It would appear that droplet sedimentation rate measurements could be made on precipitates of TMV with its whole immunoglobulin and that the use of immune Fab_1 in establishing the maximum antibody binding capacity of TMV would not be essential; the main prerequisite being that the dispersed pellet of the complex, following centrifugation, should be suspended homogeneously when the measurement is made. In this experiment 18 mg and 26 mg amounts of (anti TMV), immunoglobulin semi-purified by precipitation with 15% peg in borate buffer of pH 8.6, were reacted with 2.46 mg TMV. Similar to the previous experiments with Fab_1-anti TMV, the amount of immunoglobulin, IgG which reacted with TMV was determined in terms of BSA read from a reference line.

EXPERIMENTAL

After temperature equilibration of the contents (hydrophobic fluid) of the cylindrical glass tube a BSA concentration sedimentation rate diagram was

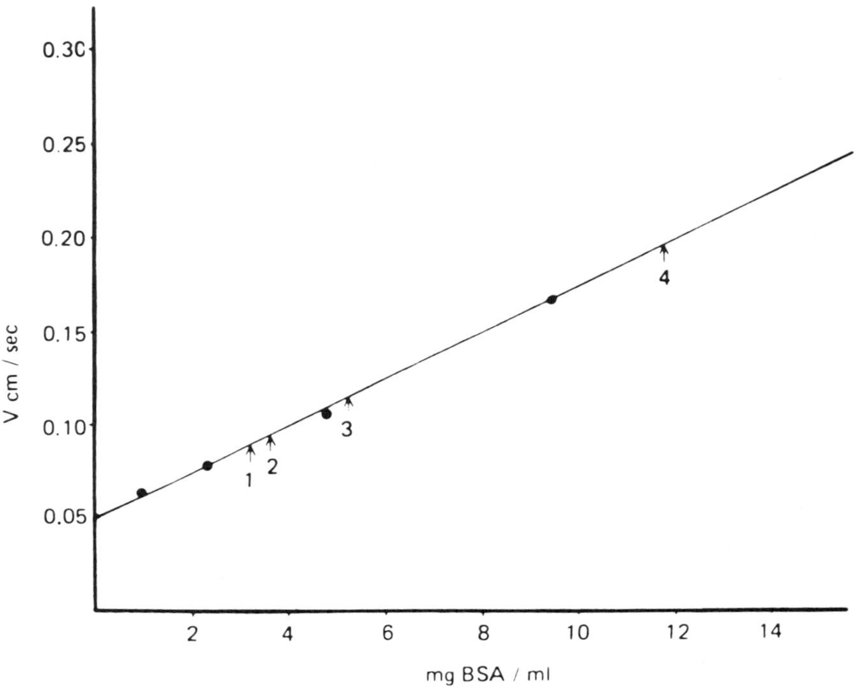

Figure 1. The relationship between concentration of bovine serum albumin in 0.1M NaC1 and sedimentation rates of 0.005 ml quantities through a hydrophobic mixture. The diagram was used in connection with the binding of immune Fab_1 to ***B.cincta*** haemocyanin and to the haemocyanin of ***H.midea***. Arrows 1 and 2 are of haemocyanin of ***H.midea*** with ***B.cincta*** immune Fab_1 and ***H.midea*** haemocyanin without Fab_1 respectively, result, no reaction. Arrows 3 and 4 are of ***B.cincta*** haemocyanin without homologous immune Fab_1 and with homologous Fab_1 respectively, result, strong reaction.

established, using twofold dilutions of the stock BSA in 0.1M NaC1. As the relationship is linear only two or three readings at different concentrations of BSA were necessary. It was essential to establish new reference diagrams at intervals, especially when several measurements were made during a period of 1 to 2h. Because 1,2-dichlorobenzene has the highest partial vapor pressure and density of the components in the hydrophobic mixture, a small amount lost by evaporation would cause a lowering of the density of the medium resulting in a faster sedimentation rate of the droplets. A sedimentation rate diagram obtained with BSA of different concentrations is presented in Figure 1.

To ensure that the available binding sites on the antigens were all occupied by Fab_1 pieces, constant masses of antigens were incubated with single and double amounts of Fab_1 before the antigen-Fab_1 complexes were ultracentrifuged into pellets. Thus two samples of an antigen, each containing x mg antigen, were mixed with y and 2y mg Fab_1 respectively. In this manner it would be possible to establish whether maximum binding of Fab_1 to the antigen had already occurred when y mg Fab_1 was used. This would be the case if the mass of the complex of antigen and Fab_1 remained constant when the mass of Fab_1 was increased to 2 y mg. The mass of Fab_1 or expressing it differently, the number of Fab_1 pieces that can bind to an antigen molecule is not a measure of the total antibody binding sites, as steric hindrance would certainly prevent some sites from being occupied by the specific Fab_1 pieces[2]. One would expect that this obstruction effect would play a major role in homologous systems such as *B.cincta* haemocyanin - anti *B.cincta* Fab_1.

Knowledge of the precise mass of antigen which would react with the specific Fab_1 was not essential, but it was attempted to use the antigen in amounts ranging from 0.3 to 1 mg. The percentages of specific Fab_1 (anti *B.cincta* and anti TMV) of the total were not known and it was for this reason that these monovalent antibody chains were used in amounts that appeared to be in excess.

The procedure followed was to add 0.1 ml of antigen (0.3 to 1 mg) to 1 and 2 ml amounts of Fab_1 solution, which had been dialyzed against 0.1M NaC1. The mixing was done in two cellulose nitrate centrifuge tubes of the No. 40 swinging bucket rotor of the Beckman preparative ultracentrifuge. After incubation for 1 hour at 22°C, the volumes of the mixture in the respective tubes were increased to 5 ml with 0.1 M NaC1. To the third tube of the swinging bucket rotor, 0.1 ml antigen and 4.9 ml 0.1M NaC1 were added. The tubes were spun for 90 min at 33,000 (approximately 120,000 x g). After centrifugation the supernatants were removed as rapidly as possible to minimize resolubilization of the pellets and liquid adhering to the walls of the tubes was removed with absorbent cotton. The three pellets were dissolved in 0.2 ml amounts of 0.1M NaC1. The rates of sedimentation of 0.005 ml aliquots of the three solutions were measured and the values obtained interpolated in a newly established reference rate sedimentation diagram of BSA. The total concentration of protein in the dissolved pellets could then be determined in terms of mg BSA/ml.

RESULTS

Evidence by Immuno-Electron Microscopy for Complete Coating of Homologous Antigens

Evidence for complete coating of *B.cincta* haemocyanin by its specific immuno Fab_1 was obtained by immuno-electron microscopy. Whereas the free haemocyanin molecules showed their characteristic short cylindrical shapes, (30 x 35 nm), formed by six layers of subunits linked together to form circles (Fig. 3), the haemocyanin molecules which were reacted with immune Fab_1 appeared to still be discrete units, although the electron micrographs indicated considerable disruption of some of the molecules (Fig. 4). The breakage possibly occurred during dispersion of the pellets. However, the complexes were larger than free haemocyanin molecules and the subunits were completely obscured and, as may be expected, the particles appeared to be more spherical than cylindrical. Electron micrographs made of different fields of the supporting electron microscope grids indicated that 100% of the haemocyanin molecules were complexed with Fab_1. Similar results were obtained with the haemocyanin of *F. varicatus* which is antigenically closely related to the haemocyanin of *B. cincta*. No complexing of anti *B.cincta* Fab_1 was noticed with the respiratory protein of *H. midea* as revealed by immuno-electron microscopy.

The difference in appearance between a TMV virion coated with immune Fab_1 and an uncoated TMV virion as revealed by electron microscopy, was not as striking as with the haemocyanin-anti haemocyanin Fab_1 system. The most obvious differences were that the central channel, characteristic of the Tobamo group of viruses, was less well defined and the surface details of the Fab_1-coated particle were different from those on the uncoated particle. No conclusion regarding increase in diameter of the TMV particle, following Fab_1 coating, could be drawn, as the edges of the particle were no longer sharply defined. Using the change in appearance of the particles as criterion, it was concluded that all the virus particles were complexed with immune Fab_1.

Binding of TMV immune Fab_1 to Y-ATMV and G-ATMV was not examined by electron microscopy in the present work.

Determination of Mass of Bound Fab_1 with Droplet Technique

The results obtained with the haemocyanins and viruses with immune Fab_1 are presented in Tables 1 and 2 and that of binding studies with undegraded immunoglobulin and TMV in Table 3 respectively.

Calculations were made of the number of Fab_1 pieces which attached to a haemocyanin molecule and to a TMV virion. The molecular weight of the haemocyanin of *B.cincta* is 8.9 x 10^6[(5)] and that of Fab_1 38,000[6]. If 1 g haemocyanin binds 1.16 g Fab_1 it follows that a haemocyanin molecule would bind 384 immune Fab_1 pieces. Similarly the number of Fab_1 pieces which bind to a TMV virion based on its molecular weight of 41 x 10^6 and that of Fab_1, can be calculated as 950.

From Table 3 it was calculated that a TMV virion could bind 442 molecules IgG on the basis that the molecular weight of IgG is 150,000. As there are two Fab_1 chains per IgG molecule, it follows that 442 intact IgG will have 884 Fab_1 units. This number should be compared with 950 Fab_1 chains calculated from the data in Table 2. Van Regenmortel and Hardie[7] found that one TMV virion could bind 400 "whole" antibody molecules, provided that both combining sites of the IgG were attached to neighboring subunits of the virus particle. This number is approximately 10% lower than the present calculation of the number of intact IgG molecules which could bind to a TMV virion.

The sedimentation diagram presented in Fig. 2 is that of hen yolk immunoglobulin and Fab_1 fragments prepared by papain digestion. This Fab_1 was used in the binding studies with the haemocyanins. Electron micrographs of uncoated haemocyanin of *B.cincta* and haemocyanin coated with immune Fab_1 are presented in Figures 3 and 4 respectively.

DISCUSSION

The sedimenting-droplet technique was applied in establishing the mass of specific Fab_1 component which could bind to 1 mg antigen. For this purpose two widely different classes of antigens were selected. One class was the haemocyanins of the molluscs *Burnupena cincta*, *Fusus varicatus* and *Haliotis midea*, all having sedimentation coefficients of 100 Svedberg units. The second class of antigens was three members of the Tobamo group of viruses viz, Tobacco Mosaic virus (TMV), Yellow Tomato Atypical Mosaic Virus (Y-TAMV) and Green Tomato Atypical Mosaic

Table 1. Binding of Fab_1-anti *B. cincta* haemocyanins obtained from different molluscs; masses of haemocyanins and Fab_1 bound given as BSA mass equivalents. Mass of Fab_1-anti *B. cincta* used 8.2 and 16.4 mg; (a) and (b) respectively.

Haemocyanin	Mass of haemocyanin	Mass of Fab_1 bound	Mass of Fab_1 bound/mg haemocyanin
B. cincta	(a) 0.540 mg	0.625 mg	1.16
	(b) 0.548 mg	0.619 mg	1.14
F. varicatus	(a) 1.24 mg	1.22 mg	0.98
H. midea	(b) 0.32 mg	0.00 mg	0.0

Table 2. Binding of Fab_1-anti TMV to virions of TMV, Y-TAMV and G-TAMV; masses of Fab_1-anti TMV used 9.3 mg (a) and 18.6 (b).

Virus	Mass of Virus	Mass of Fab_1 bound	Mass of Fab_1 bound/mg virus
TMV	(a) 0.60 mg	0.53 mg	0.88
	(b) 0.60 mg	0.55 mg	0.91
Y-TAMV	(a) 0.72 mg	0.38 mg	0.53
G-TAMV	(b) 0.42 mg	0.06 mg	0.12

Table 3. Binding of undegraded immune IgG to TMV; mass of IgG used 18(a) and 36(b) mg.

TMV mg	mg IgG bound	mg IgG bound/mg TMV
2.46	(a) 4.08	1.64
2.46	(b) 4.01	1.63
		Average 1.635

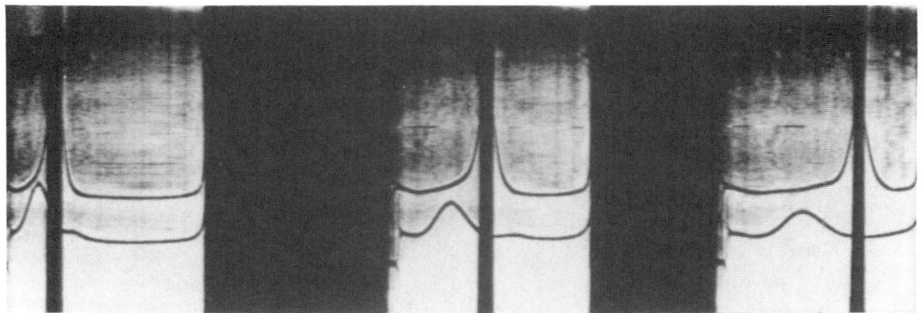

Figure 2. Analytical ultracentrifugation diagrams of intact IgY anti haemocyanin (top series) and of Fab_1 prepared from it, (lower series). The rotor velocity was 56 100 rpm, the temperature 23°C and exposures were made at 32 min intervals. Dispersion medium, 0.1M NaCl.

Virus (G-TAMV). Immune Fab_1 to *B. cincta* haemocyanin was prepared from IgY type immunoglobulin purified from the yolks of eggs laid by hens which were immunized with *B. cincta* haemocyanin[3]. Immune Fab_1 to TMV was prepared from partially purified IgG obtained from a rabbit immunized with TMV.

As could be expected, the homologous haemocyanin (*B. cincta*) bound the largest amount of immune Fab_1 per mg. The was followed by heterologous haemocyanin of *F. varicatus*. The haemocyanin of *H. midea*, which is antigenically unrelated to *B. cincta* and *F. varicatus* haemocyanins, showed no antigenic crossing with *B. cincta* and *F. varicatus* haemocyanins using *B. cincta* immune Fab_1 in the sedimenting droplet technique. Electron micrographs of *B. cincta* haemocyanin, free and complexed with Fab_1 are presented.

In droplet sedimentation experiments with three members of the Tobamo virus group, viz. Tobacco Mosaic Virus (TMV), Yellow Tomato Atypical Mosaic Virus (Y-TAMV) and Green Tomato Atypical Mosaic (G-TAMV) using TMV anti rabbit immune Fab_1, it was found that in the homologous system (TMV - Fab_1) more Fab_1 bound to TMV than to the other two members.

When considering the ratio of molecular weights of *B. cincta* haemocyanin and TMV, viz 8.9/41 = 0.22, and the ratio of the numbers of Fab_1 pieces binding to their respective surfaces, 378/950 = 0.40, it would

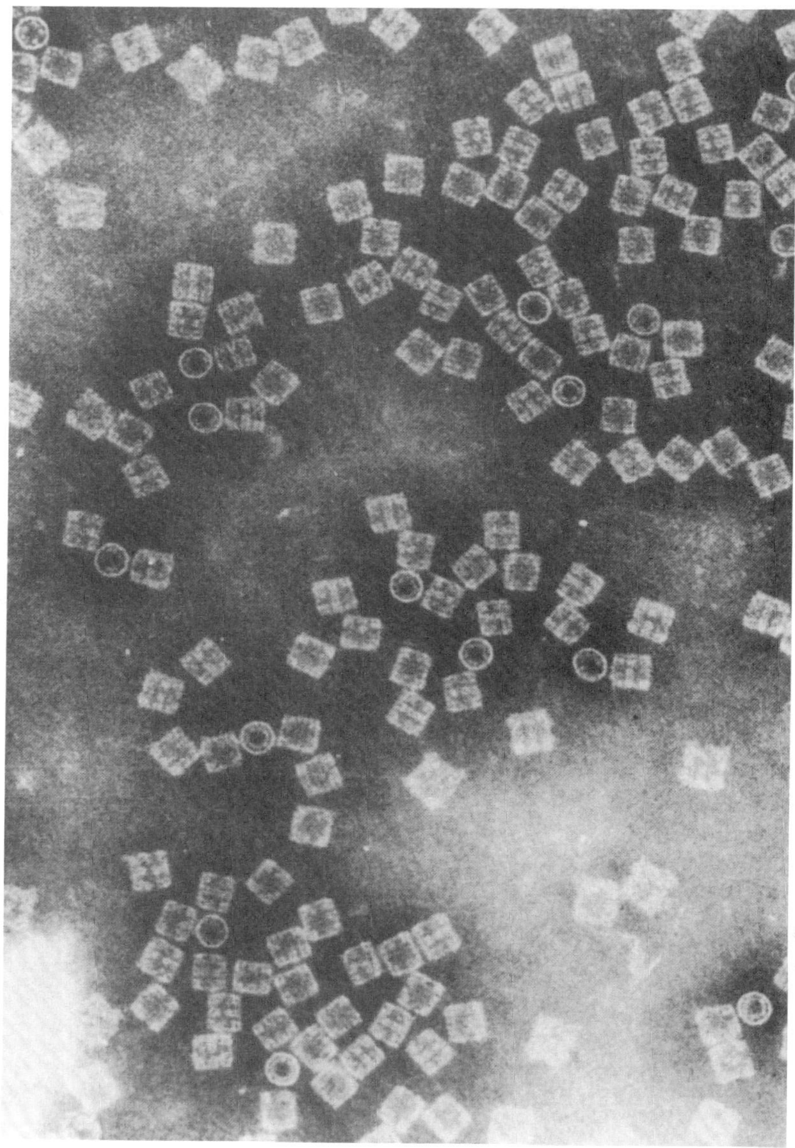

Figure 3. Electron micrograph of *B. cincta* haemocyanin negatively contrast stained with phosphotungstic acid (PTA). Length molecules 35 nm.

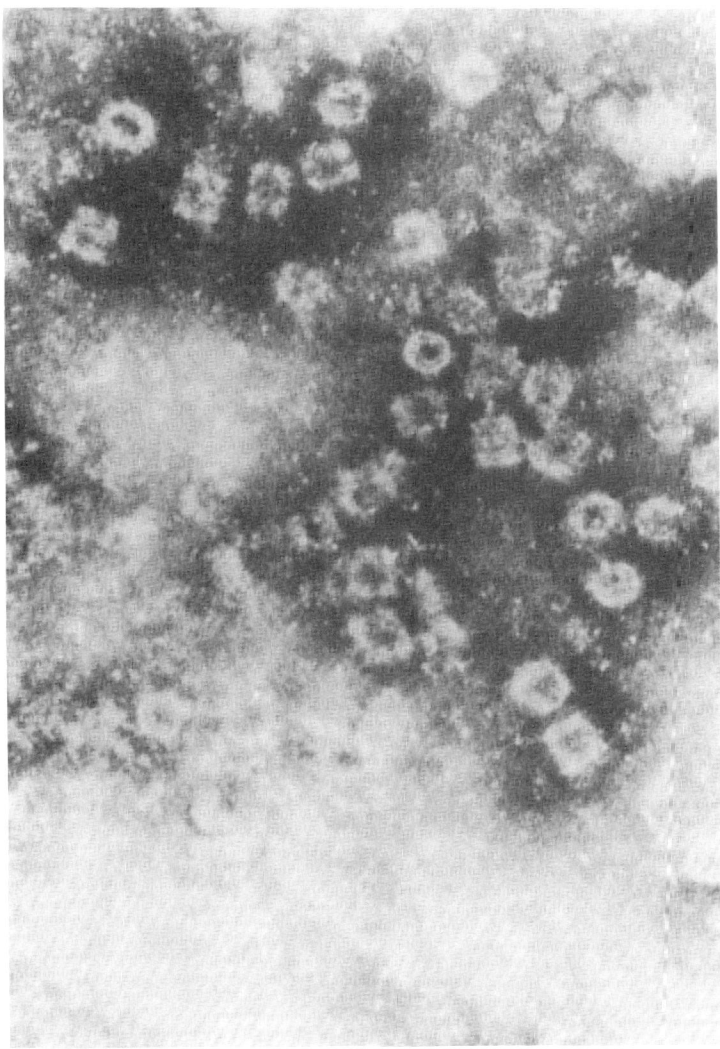

Figure 4. Electron micrograph of *B.cincta* haemocyanin reacted with immuno Fab_1. Note the irregular material on the micrograph, possibly formed by disruption during resuspension of the pellet following ultracentrifugation. Preparation, negatively contrasted stained with PTA Length haemocyanin - Fab_1 complex approximately 46 nm.

appear that the binding to the haemocyanin molecule is excessive. A possible explanation for this apparent anomaly, would be that the subunits of the haemocyanin are less tightly assembled than the subunits on the TMV virion. In this manner, relatively more surface is exposed to which Fab_1 may bind than immune Fab_1 to TMV virion. It is also possible that anti *B. cincta* haemocyanin Fab_1 may attach to antigenic sites on the interior of the haemocyanin molecules which is shown by electron microscopy to be a short cylinder, 30 x 35 nm, having an interior which is mainly waterfilled.

Van Regenmortel[2] found smaller antigenic departures of Y-TAMV and G-TAMV from TMV than may be estimated from the present communication. An explanation for the differences could be that the anti TMV serum used in the present study and that by van Regenmortel were obtained from rabbits bled at different stages following the initial injection of the antigen.

From a limited number of experiments made, it would appear that droplet sedimentation rate measurements could be made on precipitates of TMV with its whole immunoglobulin and that the use of immune Fab_1 in establishing the maximum antibodies binding capacity of TMV would not be essential; the main prerequisite being that the dispersed pellet of the complex obtained by centrifugation and dispersion in the appropriate medium, should be suspended homogenously when the measurement is made.

Although systems of whole immunoglobulins and lower molecular weight protein antigens have not been examined in the present study, no obvious reason why the sedimenting droplet technique could not be successfully applied in determining the antibody binding capacity of antigens of lower sedimentation coefficients than those of the haemocyanins and viruses used in the present study.

ACKNOWLEDGEMENTS

One of us, A. Polson gratefully acknowledges the financial support of the Medical Research Council of South Africa for this research and to the University of Stellenbosch for allowing him to conduct research on their premises.

REFERENCES

1. A. Polson, H. Alk and K. Achtleitner. Anal. Bioch. *107*, 25–31 (1980).
2. M. H. V. Van Regenmortel in Handbook of Plant Viruses. Infection and Comparative Diagnosis. E. Kurstak (Ed.) Elsevier/North-Holland Biomedical Press (1981).

3. A. Polson, Barbara M. von Wechmar and Gina Fazakerley. Imm. Comm. *9*, 495–514 (1980).
4. R. R. Porter. Bioch. J. *73*, 119–126 (1959).
5. A. Polson. Unpublished results.
6. Immuno Enzymatic Techniques. G. Feldmann, P. Druet, J. Bignon and S. Avrameas. Eds. North Holland/American Elsevier. Inserm Symposium (2) page 234 (1976).
7. M. H. V. Van Regenmortel and G. Hardie. Immunochemistry *13*, 503–507 (1976).

8. Tube Inserts as Aids in Preparative Ultracentrifugation*

ABSTRACT

Conventional preparative ultracentrifugation was modified by using inserts in the centrifuge tubes, causing a reduction of the time necessary to centrifuge a protein at reduced rotor velocity from suspension.

INTRODUCTION

Differential ultracentrifugation of viruses and glant haemocyanins may be accomplished at moderate centrifugal force in periods of 1 to 2 h. Extended periods of ultracentrifugation are required for concentrating proteins with low sedimentation coefficients using available preparative rotors and driving mechanisms. Using tubular inserts, differential ultracentrifugation is now extended to low molecular weight proteins.

Theory

During the ultracentrifugation of protein solutions or related substances, sedimentation or migration occurs in the direction of the centrifugal-force gradient. The migration of the particles ceases when they reach the bottom of a tube in the case of a swinging bucket rotor. When the wall of a tube in a fixed angle rotor is reached, the particles "slide" down the wall and concentrate at the base of the tube as a pellet or concentrate. If the

---

*From: A. Polson and K. J. van der Merwe, Prep. Biochem.: *13*, 423-436 (1983).

migration in the direction of the force gradient could be interrupted by correctly oriented solid surfaces, the sedimenting substances will form layers of higher concentration and density than the medium on such surfaces. By virtue of their higher density difference from the media, these layers will be guided by the force gradient and the correctly oriented surfaces to the lower region of the centrifuge tube. Depending on the frequency of such interfering surfaces which is inversely related to the distance between the obstacles, the rate of accumulation of the sedimenting particles at the base of the tube will vary accordingly.

METHODS AND MATERIALS

Haemocyanin was obtained from the whelk *Burnupena cincta* (*B. cincta*). This respiratory protein has a sedimentation coefficient of 100s. Crude haemocyanin (haemolymph) was obtained from the crayfish, *Jasus lalandii* (*J. lalandii*). The haemolymph consists of a main component (the haemocyanin) with a sedimentation coefficient of 16s and two minor components with sedimentation coefficients 4s and 8s, amounting to a total of approximately 15% of the protein of the haemolymph. Chicken IgY (IgG) was isolated from the yolk of eggs from normal hens according to a procedure described previously[1].

Hepatitis B antigen positive sera was obtained from the Western Province Blood Transfusion service.

Hepatitis B positive serum, free of Dane particles and consequently also free of E antigen, was a gift of Dr. Linda Stannard, Virus Research Unit, University of Cape Town.

Analytical ultracentrifugation was done in the Beckman Model E.

Inserts

Two types of inserts for use with the Beckman Corp. swinging bucket rotors were manufactured by the staff of the Department of Mechanical Engineering, University of Stellenbosch, for which the authors are most grateful. These were termed the conical cup and helical groove insert respectively. A third type of "insert" was loosely arranged polyethylene beads of average diameter 3.5 mm. Although deviating from the ideal spherical shape, these beads were found to be effective as sedimentation aids.

The various inserts are depicted in Fig. 1. From left to right are the conical cup, the helical groove and loose polyethylene bead inserts

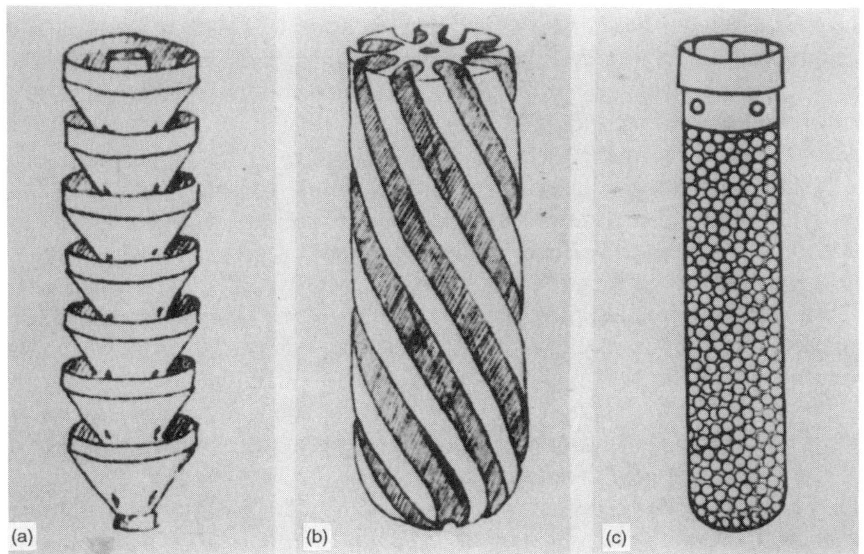

Figure 1. Centrifuge tube inserts:

(a) The Nylon conical cup insert consisting of 7 cups attached to a central rod. The cups form an angle of 37° to the vertical; 4 holes, 1 mm in diameter, drilled through each cup next to the central rod. This insert fits the tube of the SW 39 rotor.

(b) The helical groove insert milled from a rod of polyacetate; 8 grooves, 10 mm deep and 6 mm wide, cover half the circumference; distance from top to bottom surfaces of a groove, parallel to the central axis of the insert, 10 mm. This insert is used with the SW 25 rotor.

(c) A tube of the fixed angle rotor Type 65 filled with loose beads as insert. The beads may be used with the SW 39 and 25 and types 40 and 65 rotors. Length of tube 6 cm.

The conical cup insert was milled from a Nylon rod and has seven cups attached to a central rod. At the position where a cup joins the central rod, four 1 mm diameter holes were drilled. The holes allowed material to migrate during centrifugation to the cup below and eventually to the base of the tube. The surfaces of the cups form angles of 37° with the central rod. This type of insert was designed for use with the SW 39 rotor. The density of Nylon, approximately 1,15 g/cm^3, precluded the use of these inserts at centrifugal forces exceeding 58000g (22000 rpm) since at high centrifugal fields the inserts damage the bases of the centrifuge tubes.

The helical groove inserts were milled from a rod of polyacetate. The grooves were 10 mm deep and 6 mm wide. Eight such grooves, each extending over half the circumference, were cut into the rod. The width of the grooves parallel to the direction of the centrifugal force was 10 mm. These inserts were designed for the SW 25 rotor. Polyacetate is ideally suited for the purpose because of its favourable density of 0.955 g/cm^3. The consequence of its low density is that during centrifugation in an aqueous medium the insert tends to rise slightly, leaving adequate space between the insert and the lower end of the centrifuge tube to accommodate pellets of substances centrifuged from suspension.

The polyethylene beads were supplied by The Institute for Polymer Science, University of Stellenbosch. About 40% of the beads were more or less spherical and solid; the remaining 60% contained air bubbles and open cavities filled with air. This fraction was removed by flotation on a mixture of ethyl alcohol and water. The density of the beads was approximately 0.92 g/cm^3. When used, the centrifuge tubes were filled with beads to such a level that the tubes could still be capped.

Filling the tubes after the inserts had been fitted presented no problems. The tubes with the conical cup inserts were filled from the bottom upwards by inserting a syringe needle through the holes down to the base and letting the suspension in carefully. It is important that the conical cup inserts be dry, since moisture on the surfaces causes air to be trapped during filling. The tubes containing the helical groove inserts were filled from above with a pipette, and the tubes with the polyethylene beads in the normal manner in which the tubes of fixed angle rotors are filled.

After centrifugation with the conical cup inserts, fractions were removed with a syringe and truncated needle. Seven fractions, one from each cup, were obtained. It was not possible to remove fractions from different levels in the tubes with the helical groove inserts without puncturing the bases of the tubes. Fractions could be taken at different levels from the tubes with the spherical bead inserts, using a syringe and truncated needle.

The capacities of the tubes were all reduced to approximately 45% by the different inserts.

In conventional ultracentrifugation experiments on the haemocyanin of *B. cincta* at a temperature of 5°C, sedimentation free from convective disturbances could be attained in the fixed angle rotor with the 15 mm diameter tubes and in the SW 39 rotor using the 10 mm diameter tubes. When wide tubes, such as those of the SW 25 rotor were used, it was not possible to observe sharp sedimenting boundaries by means of Tyndall light-scattering by the macromolecules. The reason for this was that the horizontal rotation of the fluid, brought about by the deceleration of the rotor, eliminated any visible sedimenting boundaries. Ideal or disturbance-free sedimentation experiments could be carried out using the helical groove inserts which had been precooled to the operational temperature i.e. 4°C. The purpose of this refinement is that polyacetate has a low thermal conductivity compared to that of a metal, consequently the plastic insert will not attain the temperature prevailing in the chamber of the centrifuge rapidly. The solution in the grooves of an insert though, cools rapidly because the solution is separated from the metallic swinging bucket by a thin layer of material (i.e. nitro-cellulose) of which the centrifuge tube was made. There is therefore a temperature difference between the solution and the plastic inserts. This temperature differential will be responsible for heat convection, especially if dilute solutions were centrifuged. By precooling the insert to the operational temperature disturbances by heat convection will therefore be eliminated. When haemocyanin was centrifuged in tubes with the helical groove inserts, the sedimentation could be followed by stopping the rotor after different intervals of centrifugation.

Protein Concentration

Relative protein concentrations are expressed in terms of absorbance at 280 nm and percentages in terms of the sum of the total absorbance of the fractions. In most experiments with the bead inserts, retention of fluid by the beads was approximately 10% of the total fluid in the tube after fractionation. It is possible though, to remove the beads after centrifugation and recover all the remaining supernatant fluid (SNF) by low speed centrifugation of the beads in a separate tube.

RESULTS

The results obtained with the two haemocyanins and IgY are presented in Table 1. In these experiments the approximate time required to

Table 1. Effects of inserts on the performance of the preparative rotors SW 39, SW 25 and Types 40 and 65 fixed angle. Test proteins: *B. cincta* haemocyanin, 6 mg/ml; *J. lalandii* haemocyanin, 14 mg/ml. Percentage of total protein in the lowest 10 mm tube space T; p, pellet, t, time in min; rpm, rotor velocity. Temperature 22°C.

Insert	Rotor	rpm	Substance	S(Svedbergs)	T%	t
No insert	SW 39	30000	*B. cincta* haemocy	100	>95,p	120
Conical cup	SW 39	22000	" "	"	95,p	45
No insert	SW 25	"	" "	"	63,p	>120
Helical groove	SW 25	"	" "	"	93,p	60
No insert	Type 40	30000	" "	"	>95,p	90
Beads insert	" "	"	" "	"	95,p	10
No insert	" "	"	*J. lalandii* haemolymph	4,8,16	32 (est.)*	195
Beads insert	" "	"	"	" " "	53	195
Beads insert	" "	"	*J. lalandii* haemocy	16	90	195
No insert	Type 65	"	IgY	7	24 (est.)*	180
Beads insert	" "	"	"	7	52	180
No insert	" "	55000	"	7	32 (est.)*	180
Beads insert	" "	"	"	7	92	180

Note. The concentration in the bottom 10 mm of the tube space included the concentration of the unspun material. The estimated concentration (est.) when IgY and the haemolymph of *J. lalandii* were centrifuged without the beads, was based on the distance which the boundaries moved from the menisci.

centrifuge 50%-90% or more of the proteins into the space 10 mm from the base of the tube, either as a pellet or as a concentrate, was determined. The results obtained with the haemocyanins from *B. cincta* and *J. lalandii* as well as IgY at two different rotor velocities are presented. Diagrams of the concentrations at different levels are also given in Figs. 2, 3 and 4.

The results obtained with *J. lalandii* haemolymph are showing in Fig. 4. The upper curve is the concentration distribution following centrifugation of the undiluted haemolymph in the presence of beads at 30000 rpm (108000 x g) for 195 min. The lower curve is the protein distribution in the tube filled with beads when the concentrate in the space from the base of the tube

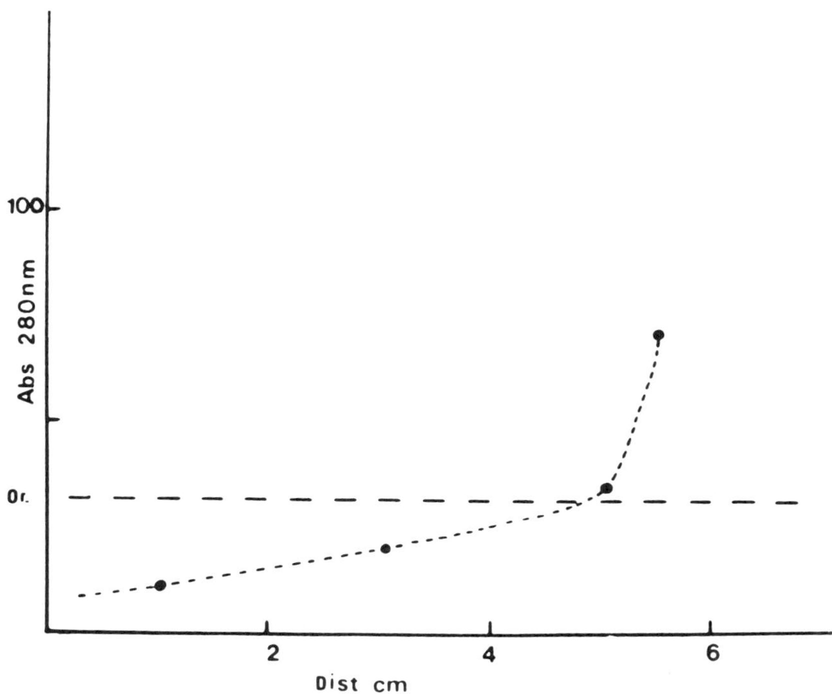

Figure 2. Distribution of IgY (chicken IgG) in a bead-filled tube of the Type 65 rotor, spun at 30000 rpm (108000 x g) for 180 min. Horizontal dashed line, original. Concentration in terms of "corrected" absorbance at 280 nm. Abscissa, distance from top of tube in cm. Medium PBS; t = 22°C.

to 10 mm above (upper curve Fig. 4) was diluted tenfold in phosphate buffered saline, pH 7.2 and recentrifuged at 30000 rpm for 195 min.

Analytical ultracentrifugal runs were made on several fractions obtained from the two preparative ultracentrifuge experiments. Details of the fractions analysed in this manner will be found in the legend to Fig. 5. Only the final ultracentrifugation diagrams are presented; the wedge and plain window cells were used. From the diagram it can be seen that the original haemolymph of *J. lalandii* is composed of one major component (representing approximately 85% of the total protein) and two minor

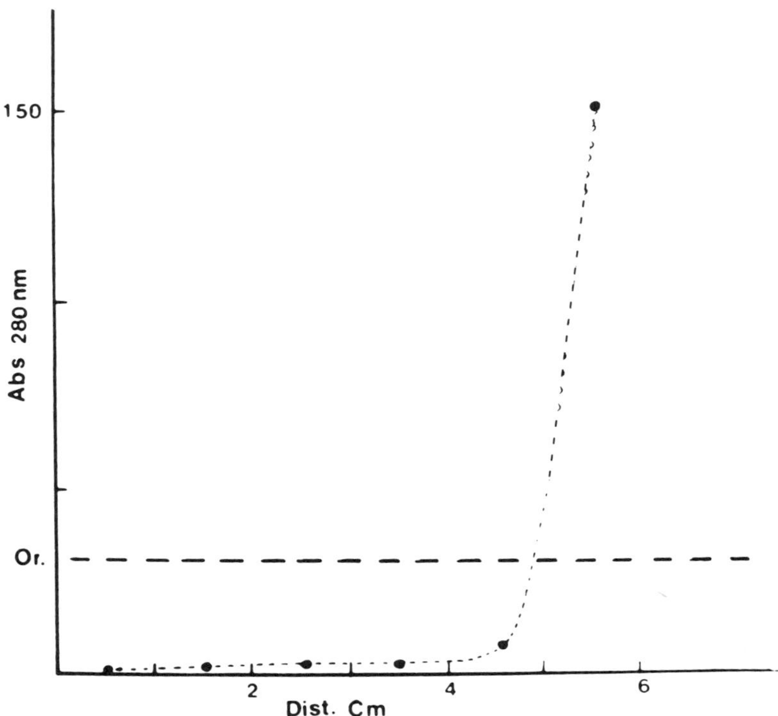

Figure 3. Distribution of IgY in a bead-filled tube, Type 65 rotor spun at 55000 rpm (302500 x g) for 180 min. Horizontal dashed line, original. Concentration in terms of "corrected" absorbance at 280 nm. Abscissa, distance from top of the tube in cm. Medium PBS; t = 22°C.

components. The sedimentation coefficient of the major component is 16.0 and the two minor entities 4.15S and 8.15S respectively. It is evident from the sedimentation diagrams that the 16S component is homogeneous and that the slower sedimenting components are enriched relative to the faster sedimenting fraction. The 16S fraction is the respiratory protein or haemocyanin of *J. lalandii*.

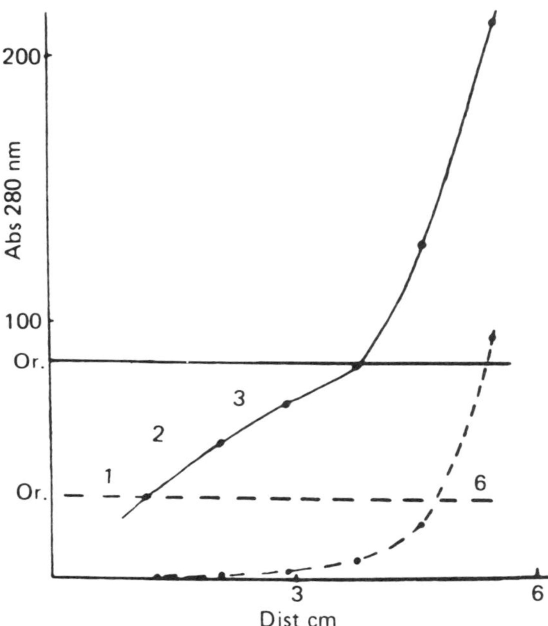

Figure 4. Upper curve, distribution of protein of undiluted haemolymph of *J. lalandii* in a bead filled tube. Type 40 rotor spun at 30000 rpm (108000 x g) for 195 min. Horizontal solid curve, original. Medium, haemolymph; t = 22°C. Lower curve (dashed), protein distribution in lead-filled tube following recentrifugation at 30000 rpm for 195 min in Type 40 rotor of the concentrate between levels 5 and 6 cm of initial centrifugation at 30000 rpm (upper curve). Horizontal line, protein content of original material. Concentration in terms of "corrected" absorbance at 280 nm. Abscissa, distance in tube in cm. Medium PBS; t = 22°C.

Hepatitis B Antigen

Hepatitis B antigen was centrifuged into a pellet using the SW 39 rotor and the polyethylene bead inserts after 60 min at 30000 rpm (108000 x g). The serum used for this purpose was shown by electron microscopy to be free of Dane particles. (Private communication, Dr. Linda Stannard.) The presence of the antigen in the pellet and its absence in the supernatant fluid

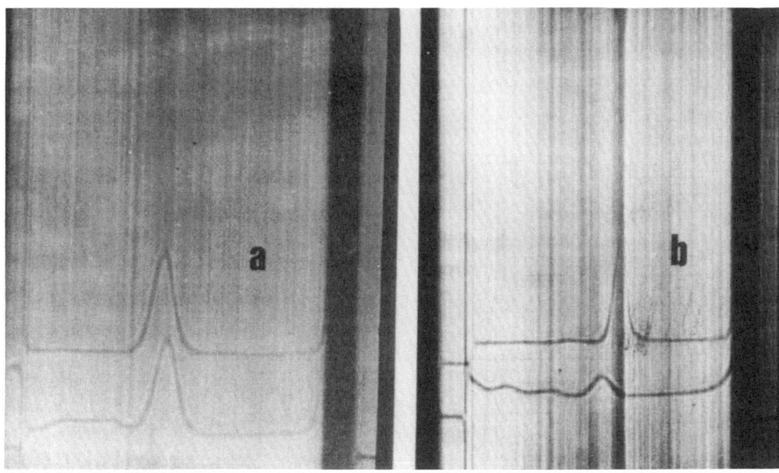

Figure 5. Analytical ultracentrifuge diagrams of *J. lalandii* haemolymph and fraction. Frame a, lower diagram, original haemolymph diluted 4 x in 0.15 M NaCl. Top diagram, the haemocyanin centrifuged down twice at 108000 x g for 195 min., see lower curve Fig. 4, fraction diluted 10 x in 0.15 M NaCl before analytical centrifugation. 38000 rpm time approximately 90 min.

Frame b, lower diagram, pool fractions 1.2 and 3 and diluted 4 x in 0.15 NaCl before analytical ultracentrifugation. Top diagram fraction 6, upper curve Fig. 4, diluted 10 x before analytical ultracentrifugation. 48000 rpm, time approximately 90 min.

was shown by Ouchterlony gel diffusion against Hepatitis B antigen IgY type antibody. The Hepatitis B antigen was also spun from three of four sera obtained from the Western Province Blood Transfusion Service. The presence in the pellets and its absence in the SNF was confirmed by the Ouchterlony test.

DISCUSSION AND CONCLUSIONS

Several inserts for preparative centrifuge tubes were used in an attempt to reduce the time required to centrifuge macro-molecular weight substances from solution either in pellet form, or as concentrated solutions. The function of the inserts was to provide obstructing surfaces for the sedimenting particles. The particles formed layers of higher concentration or densities on these surfaces. These surfaces were oriented in such a manner that the fluid of higher density would be directed by the centrifugal force gradient towards the lower regions of the centrifuge tubes. The substance thus centrifuged from suspension was collected at the bases of the tubes either as pellets or as concentrated solutions.

Three types of inserts were used of which two were designed for use with the swinging bucket rotors of Beckman preparative centrifuges. The first two of these inserts were termed conical cup and helical groove inserts respectively. The third insert was an assortment of near spherical beads of polyethylene with an average diameter of 3.5 mm. The beads were used in conjunction with both the swinging bucket and fixed angle rotors. Perfectly spherical beads would possibly have performed better as inserts, but were not available. It is uncertain whether the average diameter of 3.5 mm is optimal, beads with a smaller average diameter, for example 2.0 mm, may be more suitable.

The conical cup inserts were milled from a Nylon rod as it appeared to be the plastic material most suitable for manufacturing the delicate structures of this type of insert. Nylon also has the desired tensile strength which enables it to resist high centrifugal stresses. Unfortunately its density, which is higher than that of water, precluded its use at rotor velocities higher than 22000 rpm (58000 x g) because the inserts damage the bases of tubes at high velocities.

The helical groove inserts were milled from a rod of polyacetate. This material has a density of 0.955 g/cm^3 which is ideal for inserts. Unfortunately polyacetate cannot be milled as finely as Nylon.

The "near" spherical beads of polyethylene have a density of 0.922 g/cm^3 and when packed into either the swinging bucket or the fixed angle rotors provide the short distances between the "obstructions" necessary for rapid concentration.

From Table 1 it is evident that the bead inserts were the most effective as preparative ultracentrifugation aids. The time required to centrifuge the haemocyanin of *B. cincta* into a pellet at a rotor velocity of 30000 rpm using the Type 40 rotor without the beads was 90 min. With beads the time was reduced to 10 min. With the beads in the tube, the conical cup and helical groove inserts were less effective, although g x t, was reduced markedly. The factor g is the centrifugal force and t the duration of centrifuging the haemocyanin of *B. cincta* into a pellet.

The inserts reduced the capacity of the centrifuge tubes by 55% but this unavoidable disadvantage is offset by the advantage of rapid sedimentation.

By using beads as inserts and chicken IgG (IgY) as test substance, it was possible to centrifuge the protein out of solution at 55000 rpm with the Beckman Type 65 rotor in 180 min. and the haemocyanin of *J. lalandii* was separated as a single component with a sedimentation coefficient of 16S from associated proteins by two steps of differential ultracentrifugation at 30000 rpm (108000 x g).

Hepatitis B antigen could be centrifuged at 30000 rpm (108000 x g) from suspension in 60 min.

In the present study centrifugal forces of 300000 x g in conjunction with bead inserts have shown to be of value for centrifuging substances with sedimentation coefficients of 7S from solution. Since modern preparative ultracentrifuges generate higher centrifugal forces, differential centrifugation of proteins of lower S values on a preparative scale is now a distinct possibility.

In conclusion it may be stated that using the bead inserts, the time necessary to centrifuge a component from suspension to the bottom of a centrifuge tube does not appear to be a linear inverse function of the sedimentation coefficient. The most likely explanation for this behaviour seems to be that reverse diffusion is the factor responsible for the non-linear relationship.

ACKNOWLEDGEMENTS

This research was supported by funds supplied by the Medical Research Council of South Africa, the South African Inventions Development Corporation and Fine Chemicals, Cape Town. One of us, AP is grateful to the University of Stellenbosch for use of their premises.

REFERENCE

1. A. Polson and M. Barbara von Wechmar, Immunological Communications, 9(5), 475-493 (1980).

9. Bead Centrifuge Tube Inserts — A Semi-Quantitative Theory*

ABSTRACT

A semi-quantitative theory is proposed for the sedimentation of particles in centrifuge tubes filled with polymer beads acting as sedimentation aids. The theory concerns the attainment of approximate sedimentation-diffusion equilibrium in minimum time of centrifugation. It is shown empirically that the time to reach this state is approximately proportional to the diameter of the polymer beads as predicted by the theory.

INTRODUCTION

In an earlier communication to this Journal[1] the use of centrifuge tube inserts, for reducing the time required for centrifuging a protein from solution was described. It was stated that, apart from sedimentation, diffusion in the reverse direction plays an important role in the rate of concentrating a substance through the use of the centrifugation aids. The most effective aid appeared to be spherical beads of uniform diameter (≤ 3 mm). In the following report it is attempted to formulate a theory from which the minimum time of centrifugation necessary for centrifuging the bulk of a substance into a defined space at the base of the centrifuge tube, may be predicted.

From: A. Polson and K. J. van der Merwe, Prep. Biochem., *14*: 223-230 (1984).

THEORY

Consider a centrifuge tube filled with beads of uniform diameter. Assume, for the lack of definition, that the empty spaces amongst the beads have an average length Δ x cm in the direction of the axis of the tube. The number n, of such spaces parallel to a line extending from the upper surfaces of the beads in the tube to the base of the tube will then be:

$$n = \frac{x}{\Delta x} \tag{1}$$

in which x is the length of the bead column.

If a solution of a protein with sedimentation and diffusion coefficients S and D respectively, were centrifuged in the tube filled with beads of uniform diameter, the transfer of material M_s, at any level r, by the centrifugal force, in time dt would be:

$$M_s.dt = n\ S\ c\omega^2 r_x\ (1 - \bar{v}\rho m)\ dt \tag{2}$$

where n, the number of spaces of length Δx,

c, the concentration
ω, the angular velocity of the rotor,
r_x, the distance of the level from the centre of rotation,
$\bar{v}$, the partial specific volume of the substances and
ρ_m, the density of the medium.

The average value of $\bar{v}$ for proteins is 0.737 g/cc.

As sedimentation occurs throughout the tube, regions of higher concentration are formed closer to the surfaces of the beads than away from the beads. Because these regions have higher densities they will move in the direction of the centrifugal force at a faster rate than the protein would have done if the interfering beads had not been present. The downward migration of the concentrates is opposed by reverse diffusion of the protein. The opposing transfer of protein is described by Fick's first law.

$$M_D.dt = D\frac{dc}{dx}\ .dt \tag{3}$$

in which M_D is the transport of protein at level rx and dc/dx is the concentration gradient at the same level. Assume that approximate sedimentation equilibrium is reached when 95% or more of the total protein in the tube had sedimented into the region of the tube between the base and 1 cm above the base. Regarding the position 1 cm above the base as r_x, the net transfer across this level is zero because the transfer by sedimentation equals that by diffusion in the reverse direction.

Thus $M_D = M_S$ or

$$n\ S\ c\ \omega^2\ r_x\ (1 - \bar{v}\rho_m) = D\frac{dc}{dx} \tag{4}$$

Upon rearrangement follows

$$n\frac{S}{D}\ c\ \omega^2\ r_x\ (1 - \bar{v}\rho_m) = \frac{dc}{dx} \tag{5}$$

Regarding $n\ (1 - \bar{v}\rho_m) = K =$ a constant for a defined bead size, we may write

$$n\ (1 - \bar{v}\rho_m) = K = \text{a constant}$$

$$K\ \omega^2\ r_x\frac{S}{D}\ .dx = \frac{dc}{d} \tag{6}$$

After integration between the limits r_x and $r_x + 1$ follows,

$$K\ \omega^2\ r_x\frac{S}{D}\ [(r_x + 1) - r_x].\ = \text{in}\ \frac{c_1}{c_2}\ \text{or,}$$

$$K\ \omega^2\ r_x\frac{S}{D} = \text{in}\ \frac{c}{c_2} \tag{7}$$

in which c_2 is the "average" concentration below level r_x and C_1 that at level r_x. It is not possible to use the value of the concentration at the base as from the theory this should be infinitely high. The nearest that this state could be approached to is when substances of high molecular weights ($M_r \geq 5 \times 10^5$) such as haemocyanins and viruses are spun into pellets. Proteins of lower molecular weights like IgG and serum albumin form semi-solid or viscous masses when spun from solution. The thicknesses of their layers on the bases of the centrifuge tubes are dependent upon the concentration of the unspun or original material.

Regarding $C_2/C_1 = 95$ as an arbitrary index for approximate sedimentation - diffusion equilibrium, the conditions of ultracentrifugation may be established empirically for a substance of known parameters S and D, and to use such data for calculation of the conditions of centrifugation for other proteins from their known or estimated S and D values. Such data for proteins of widely different molecular weights are given by Svedberg and Pedersen[2].

Table 1. Calculated and observed time of ultracentrifugation for concentrating different proteins in the lowest region of a tube filled with 3 mm beads, Type 65 rotor; medium, 0.1M phosphate buffer, pH 7.2; T, 20°C. Volume of protein centrifuged 5 ml. D expressed in 10^{-7} cm^2/sec.

Substance	S	D	$\omega^2 r_x$(g)	t min calc.	t min obs.
Haemocyanin (*J. lalandii*)	16.4	3.4	108000		195
Haemocyanin (*Burnupena cincta*)	100	1.2	108000	11	10
IgG (human)	6.5	4.6	302000	200	180
Serum albumin (human)	4.3	6.1	302000	400	360

MATERIALS AND METHODS

The haemocyanin of *Jasus lalandii* (*J. lalandii*) is ideal as reference protein because of its stability in solution, its sedimentation and diffusion coefficients, (which are intermediate for globular proteins) and its blue-grey colour in solution which enables one to follow its sedimenting boundary in the tubes filled with beads, during ultracentrifugation.

Polystyrene beads of 3 mm and 1 mm diameter were kindly supplied by the Institute of Polymer Research, University of Stellenbosch.

RESULTS

In Table 1 are given data calculated and obtained empirically for several proteins, using known values of S and D for *J. lalandii* and the time required for centrifuging the protein from solution at 20°C. The parameters S and D were determined previously and found to be identical to those of the haemocyanin of the closely related species, *Palinurus vulgaris*[2, 3].

The Beckman Type 65 rotor was used at 20°C and the beads used had a diameter of approximately 3 mm. Relative protein concentrations were obtained from uv absorbance measurements at 280 nm.

The agreement between the calculated and observed values of t, expressed in minutes of centrifugation, is satisfactory considering that the temperature of the rotor may not be constant during centrifugation. A higher temperature than 20°C would result in a lower viscosity of the medium and a faster rate of sedimentation. As a result of the high concentration in the lowers "1 cm" regions of the centrifuge tubes anomalous behaviour of the material towards the centrifugal forces may also be expected. The rate of increase in concentration between the base and 1 cm above the base, is expected to be a linear function of time during the initial stages of centrifugation but, when the state of equilibrium is approached there would be a steep concentration gradient dc/dx in the defined space of the tube and transport by diffusion $M_D = D\frac{dc}{dx}$, in the reverse direction would decrease the rate of concentration in the space.

The theory predicts that by decreasing the size of the beads, n will be larger, and faster attainment of the desired concentration in the defined space would occur.

To verify the approximate validity of equation (4) in respect to the factor n and to the assumption of approximate sedimentation-diffusion equilibrium, an experiment was designed and conducted on a solution of 15 mg/ml *J. lalandii* haemocyanin in 0.1 M phosphate buffer pH 7.2. The respiratory protein was spun in a tube filled with polystyrene beads of 1 mm average diameter; the Type 65 fixed angle rotor of a Beckman preparative ultracentrifuge was used and the temperature was maintained at 20°C. The rotor was spun at 108000xg (30 000 rpm) for a total of 80 min during which period the centrifuge was stopped at 20 min intervals for visual inspection of the extent of migration of the sedimenting boundary. After 60 min the boundary was approximately 1.5 cm from the base of the tube and after 80 min the bulk of the respiratory protein was contained in the space between the base and 1 cm above the base. From this observation it was concluded that the boundary had sedimented into the "1 cm" space in approximately 70 min (average 60-80 min) in the presence of the beads of 1 mm diameter. When compared with the result obtained with the "3 mm" beads viz. 195 min, (Table 1), it would appear that the use of the parameter n, in its linear form is correct (equation 4).

To indicate that approximate equilibrium was reached after centrifugation for 70 min in the present experiment, centrifugation of the tube with the haemocyanin in the lowest 1 cm space was continued for an additional 150 min. On examining the tube, the position of the boundary at the end of this additional period was still at the 1 cm level above the base, indicating that approximate sedimentation diffusion equilibrium had been reached after 70 min centrifugation at 108000xg.

It was not possible to remove fractions at different levels in the tube filled with the 1 mm beads for assay as was done following centrifugation with the 3 mm beads[1]. To cover the haemocyanin concentrate, the tube was held vertically for 16 hr, allowing the loose pellet to redissolve, piercing the bottom of the tube with a syringe needle and collecting the droplets of the concentrated haemocyanin.

DISCUSSION AND CONCLUSIONS

It would appear that by observing the principles of equilibrium ultracentrifugation of Svedberg[2] to meet the requirements of sedimentation in tubes filled with polymer beads of uniform size, the time required to centrifuge a protein from solution may be predicted using the parameters established for a reference substance.

The ability to centrifuge any protein from solution with the aid of tiny beads of polymer, raises the possibility of using solid ion exchange particles of uniform diameter in preparative ultracentrifugation. Thus, it should be possible to retain a component on the beads while the unattached components are removed by the centrifugal field. By changing the ionic strength or pH, the bound component may then be centrifuged off into a concentrated state or even a pellet.

ACKNOWLEDGMENTS

The authors are grateful for financial aid received from Messrs. Fine Chemicals, Epping, Cape Town, the South African Inventions Development Corporation and the Medical Research Council of South Africa and to Mr. P. Swart of our Department for valuable discussions. One of us, A. P. would like to thank the University of Stellenbosch for the facilities provided for conducting this research.

REFERENCES

1. A. Polsen and K.J. van der Merwe, Prep. Biochem. *63*, 423-436 (1983).
2. T. Svedberg and K.O. Pedersen. The Ultracentrifuge. Oxford at the Claredon Press, Oxford, Page 406 (1940).
3. T. Svedberg and K.O. Pedersen. The Ultracentrifuge. Johnson Reprint Corp., New York (1959).

10. Purification of Viruses by Centrifugation in Gradients of Inert Substances*

A. Polson
Infertility and Reproductive Biology Unit
Department of Obstetrics and Gynecology
Medical School, University of Stellenbosch
Tygerberg Hospital, Tygerberg
Republic of South Africa

ABSTRACT

By using a reorienting gradient centrifuge rotor cut from a block of Nylon and fitted with eight septae, it was possible to separate the components of the haemolymph of the mollusc *Turbo sarmaticus* into three fractions in a sucrose gradient held in the bowl of the rotor. The fractions were (108 and 98)S, 44S and 16–22S. The success of the experiment was due to the large differences in the sedimentation coefficients of the components. When the rotor was applied to the natural mixture of the five viruses of the caterpillars of *Nudaurelia cytheria* only the main component could be isolated in a pure state. The viruses were separated by isopycnic centrifugation in "self formed" caesium chloride gradients, using a Beckman Model E analytical centrifuge in which a separation cell fitted with a centerpiece with two perforated partitions was used.

*Hitherto unpublished.

INTRODUCTION

Centrifugation in gradients of inert substances is useful for the separation of components in a mixture[1]. There are two principles involved in this type of separation. One, termed reorienting gradient centrifugation (reograd) relies on the differences in masses or, better still on the sedimentation coefficients of the different components in the mixture and the second, termed isopycnic centrifugation[2], on the densities or specific gravities of the different entities.

METHODS AND MATERIALS

Reorienting Gradient Centrifugation

Substances of high molecular weight such as the components in the haemolymphs of molluscs the haemocyanins, or mixtures of viruses or viruses contaminated with extraneous matter may be separated from one another or from extraneous matter by centrifuging in preformed gradients of inert substances. In this type of separation also termed rate zonal separation, the optimal time and rotor velocity is determined in a preliminary run. The centrifugation may be done in gradients of sucrose or glycerine preformed prior to the insertion of the mixture to be fractionated. The material is layered on the gradient when the rotor is at rest and the sucrose or glycerine has a vertical gradient or, the mixture may be introduced during spinning of the rotor. Details of some of the rotors used may be found in a brochure obtainable from the Beckman Corporation[2].

In the present communication results obtained with a rotor cut from a block of Nylon and furnished with eight septae radiating from a central cup are given. The function of the cup is to hold the material to be fractionated during acceleration. When a certain velocity is reached the fluid spills over the brim of the cup on the gradient from where the different components of the mixture sediment outwards as zones. On acceleration, the gradient changes from the horizontal to the vertical. The gradient is fractionated by syphoning the contents of the bowl through a long truncated needle into a series of calibrated tubes. The U.V. absorbances at 280 nm of the different aliquouts are plotted against their number.

An important aspect of the Nylon reograd rotor is that the diameter of the central cup from where the sample is introduced is such that the material is layered on the gradient where the centrifugal force is already

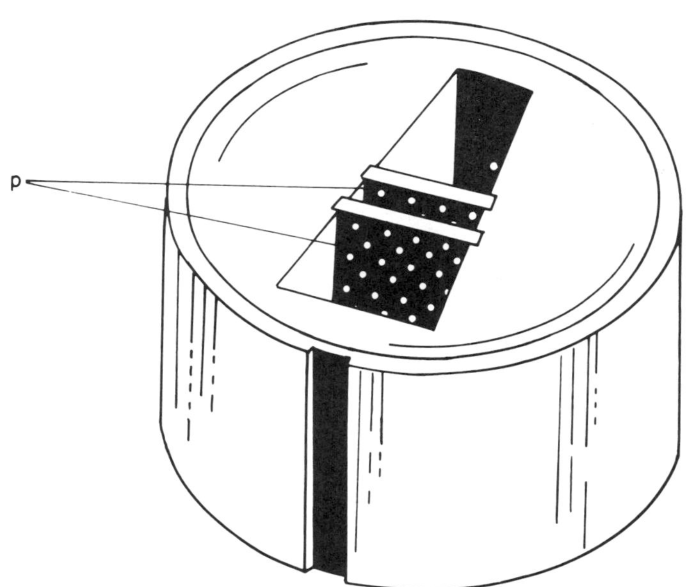

Figure 1. Cell centerpiece of the analytical centrifuge rotor showing the perforated plastic partitions which divide the cavity into three compartments. The partitions are 2 mm apart.

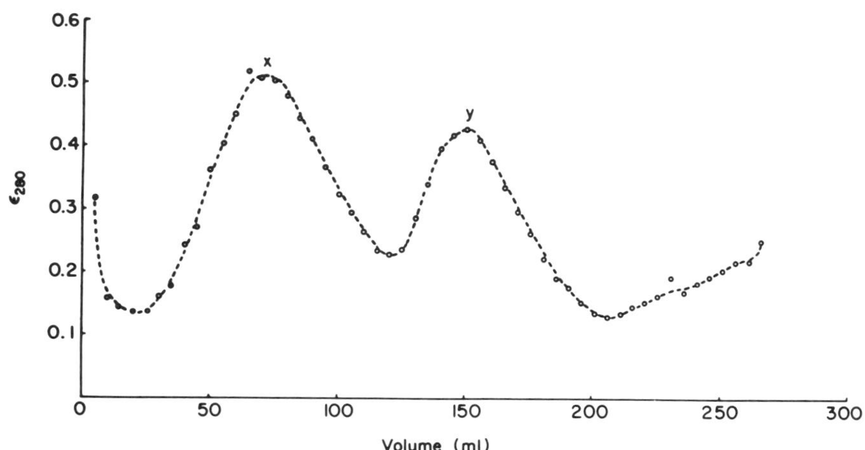

Figure 2. Ultraviolet absorption diagram at 280 nm of the contents of the nylon reograd rotor after centrifugation of the haemolymph (haemocyanin) of *Turbo sarmaticus* in a sucrose gradient at 12500 rpm for 18 hours at 21°C; phosphate buffer 100 mM pH 7.2.

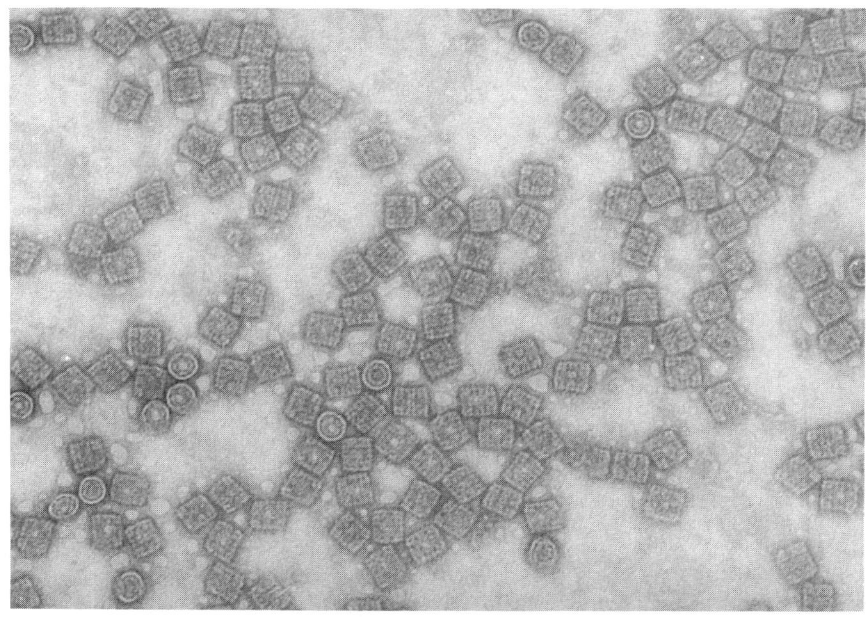

Figure 3. Electron micrographs of the two components, X, (S108 and S98) and Y, S44. X and Y of Figure 2.

approximately 50% of that on the outer wall of the rotor. A detailed description of the rotor may be found in two publications[3,4].

Isopycnic Gradient Centrifugation

In isopycnic gradient centrifugation the mixture to be fractionated is dissolved in a solution of a high density salt which is inert such as CsCl or Cs_2SO_4. When the mixture of salt and material is spun at high rotor velocity (35,000 to 56,000 rpm) in a Beckman preparative centrifuge for a period of 12 to 18 hours, the high density salt solution forms a sedimentation diffusion equilibrium gradient. In this gradient the different components of the material to be fractionated condense at levels in the tube equivalent to their own densities. After completion of the isopycnic run the components are obtained by the method known as bottom puncture.

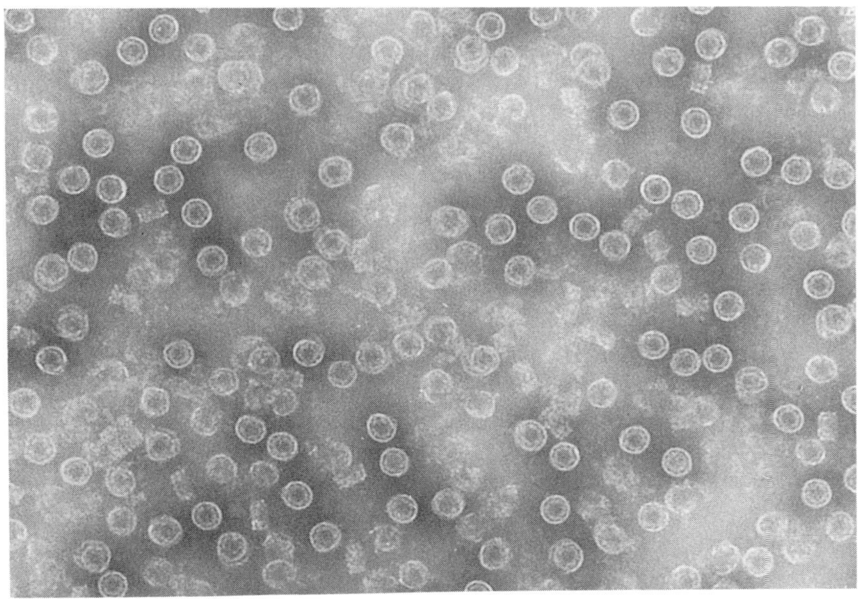

Figure 3. Continued

Isopycnic gradient centrifugation may also be conducted in the Model E analytical centrifuge in which the separation cell is used. The centerpiece of the separation cell is provided with a single perforated partition near the central region of the cavity. When a mixture is centrifuged to isopycnic equilibrium two fractions may be obtained, a fraction which condenses below the disc and one above. The fractions may then be removed from the cavities with a syringe fitted with a 20 gauge needle.

In the present work a centerpiece, (Figure 1) fitted with two perforated partitions was used[5]; this centerpiece was made to our design by Beckman Corp. When a mixture of a virus and extraneous material or a mixture of viruses is centrifuged to isopycnicity, the components may be made to condense above, between and below the partitions by selection of the appropriate density of the Cs salt solution and rotor velocity. An advantage of this method of centrifugation is that the gradient and the positions of condensation of the different components may be followed with the telescope of the centrifuge and if necessary the gradient of the Cs salt solution in the centerpiece may be altered by increasing or decreasing the rotor velocity until the most favorable distribution for isolation of the components in the cell is obtained.

RESULTS

Reorienting Gradient (Reograd) Centrifugation

A reograd sedimentation diagram obtained of the haemolymph of the molluscs *Turbo sarmaticus* (Ts) is shown in Figure 2. The diagram was obtained when the haemolymph was spun in a sucrose gradient at 12500 rpm for 18 hours at 21 °C, details are given in the legend to Figure 2. Electron micrographs (Figure 3) X and Y were made of the fractions represented by the two peaks X and Y. The two haemocyanins, 108 and 98S, could not be differentiated on the electron micrograph but the Y peak was clearly of the 44S entity.

The Nylon reograd rotor was also applied to a natural mixture of Nudaurelia moth (N) viruses. Upon centrifugation at 12500 rpm for 9 hours at 21 °C in a similar sucrose gradient as that used with the *Ts* haemocyanin, the UV absorption diagram at 280 nm depicted in Figure 4 was obtained. Electron micrographs made of material represented by positions C, B and A indicated that material from fraction B was homogeneous in particle size. (Figure 5); C showed a mixture of two viruses with nearly the same diameters and fraction A was a mixture of a small and a larger component. The reason for the limited success may be due to the similarity in sedimentation coefficients of the viruses. It may be stated that the diagram obtained with the reograd rotor bears a striking resemblance to the sedimentation diagram obtained with the Model E analytical ultracentrifuge.

Isopycnic Centrifugation in the Model E Analytical Centrifuge

The following isopycnic ultracentrifugation experiments were performed on a purified natural mixture of the viruses of the emperor moth *Nudaurelia cytheria cytheria* (N):

(a) An isopycnic run in a CsCl gradient in a cell with a centerpiece without a perforated partition. The rotor velocity was 50400 rpm, the time of centrifugation 5 hours. The initial density of the CsCl solution was 1.310g/ml and the temperature 22 °C.

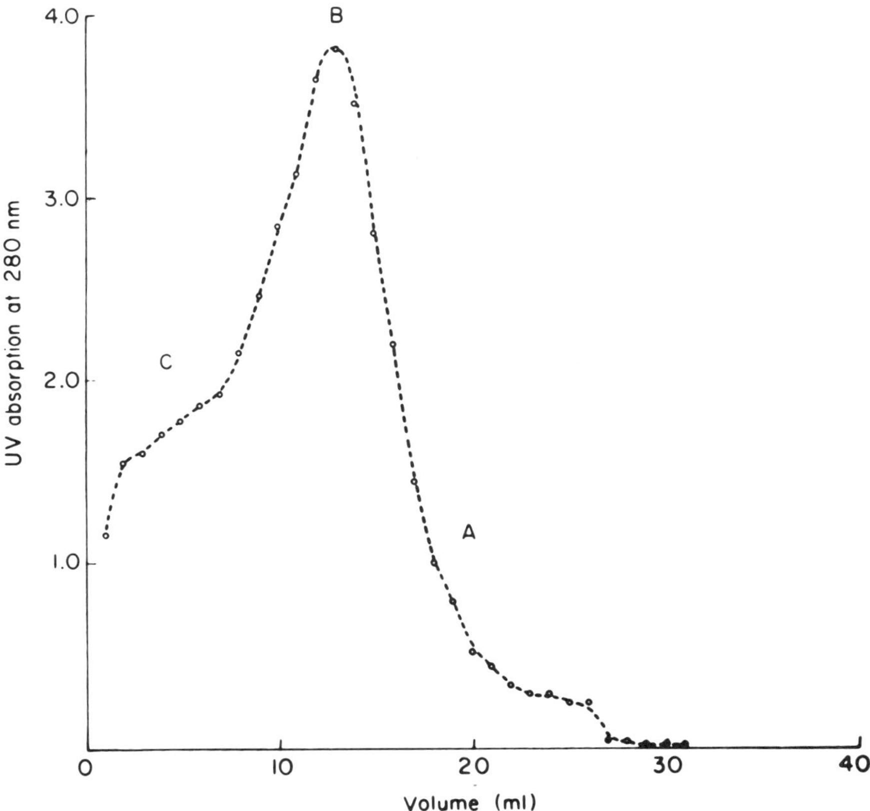

Figure 4. UV absorption diagram at 280 nm of a mixture of *Nudaurelia* viruses obtained by centrifugation in a sucrose gradient at 12500 rpm at 21°C for 12 hours in the reograd rotor. The fractions A, B and C were subjected to electron microscopy.

(b) An isopycnic run in a CsCl gradient in which the double partition centerpiece was used at 50400 rpm for 5 hours at 22°C. The initial density of the CsCl was 1.310g/ml.

(c) An isopycnic run in the cell with the double partition centerpiece. The rotor velocity was 44700 rpm and the time of centrifugation 5 hours at 22°C. The initial density of the CsCl was 1.340g/ml.

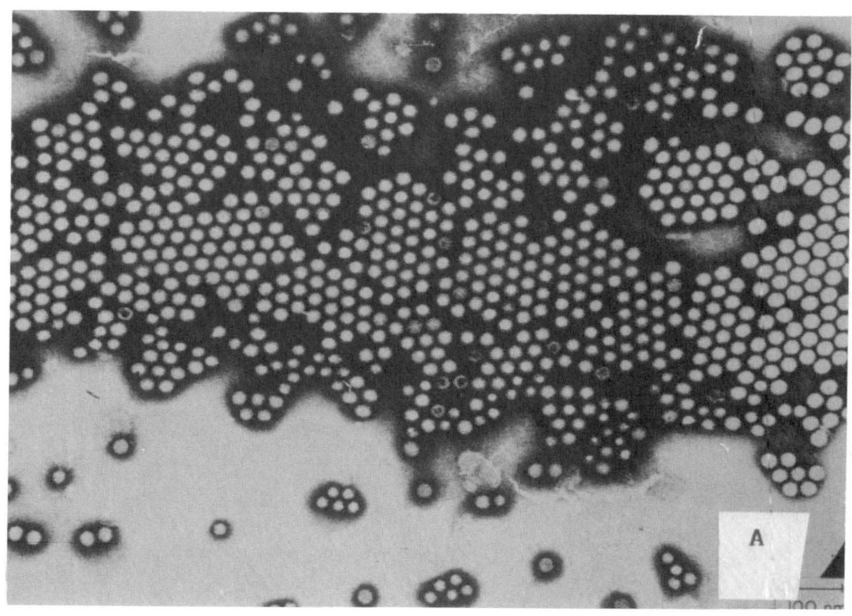

Figure 5. Electron micrograph of material from fractions C, B and A (Fig. 4). Fraction C is a mixture of β and ϵ, fraction B pure β and fraction A a mixture of β and γ Nudaurelia viruses respectively.

The following isopycnic centrifugation diagrams were obtained (refer above):

(a_1) Figure 6, an isopycnic diagram of the N viruses in a cell without partitions,

(b_1) Figure 7, an isopycnic centrifugation diagram of the N viruses in a cell with two partitions, and

(c_1) Figure 8, an isopycnic centrifugation diagram with two partitions in the centerpiece. Details are in legends to Figures.

The component between the partitions, diagram (b_1) was isolated with a syringe and a 20 gauge needle, freed of CsCl and subjected to electron microscopy (Figure 9). The agent was identified as Nβ virus and the

Figure 5. Continued

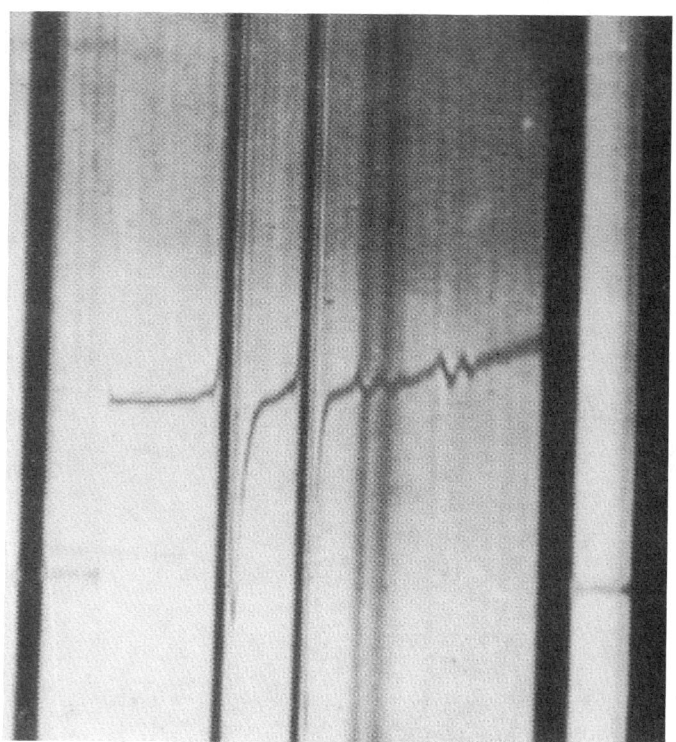

Figure 6. Isopycnic ultracentrifugation diagram of purified N viruses after centrifugation for 5 hours at 50400 rpm in a "self formed" gradient of CsCl solution; initial density of the CsCl solution 1.310 g/ml. The centerpiece of the cell was without the perforated partition. Note the resolution of the components into six entities, one indicated by the opalescent band is possibly of denatured proteins.

component immediately below the second partition as the Nϵ virus. The component between the partitions (c_1) was identified as NΓ virus and the components near the base and above the first partition the remaining two Nα and Nδ; the separate viruses were not identified.

DISCUSSION

Two methods of separating and purifying viruses by centrifugation in density gradients of inert substances are described briefly, the methods are

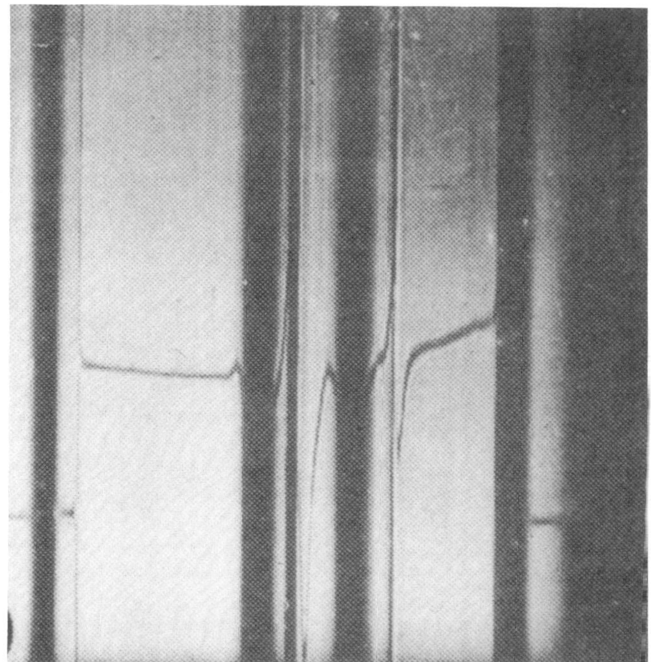

Figure 7. Isopycnic centrifugation diagram in a "self formed" CsCl gradient of purified N viruses using the cell with the double perforated partition centerpiece. Note the condensation of one component in the cavity between the two partitions. The two major components Nβ and Nδ were isolated in a pure state. Initial density of the CsCl 1.310 g/ml; rotor velocity 50400 rmp; time of centrifugation 5 hours. The change of the density gradient due to the presence of the two perforated partitions.

reorienting gradient (reograd) and isopycnic centrifugation. Rate zonal centrifugation experiments were conducted in a reorienting gradient (reograd) rotor cut from a block of Nylon in our workshop and the isopycnic centrifugation experiments were conducted in the Model E analytical ultracentrifuge containing a cell with a centerpiece fitted with two perforated partitions made by Beckman Corp. to our design. Caesium chloride was used as the self forming gradient material.

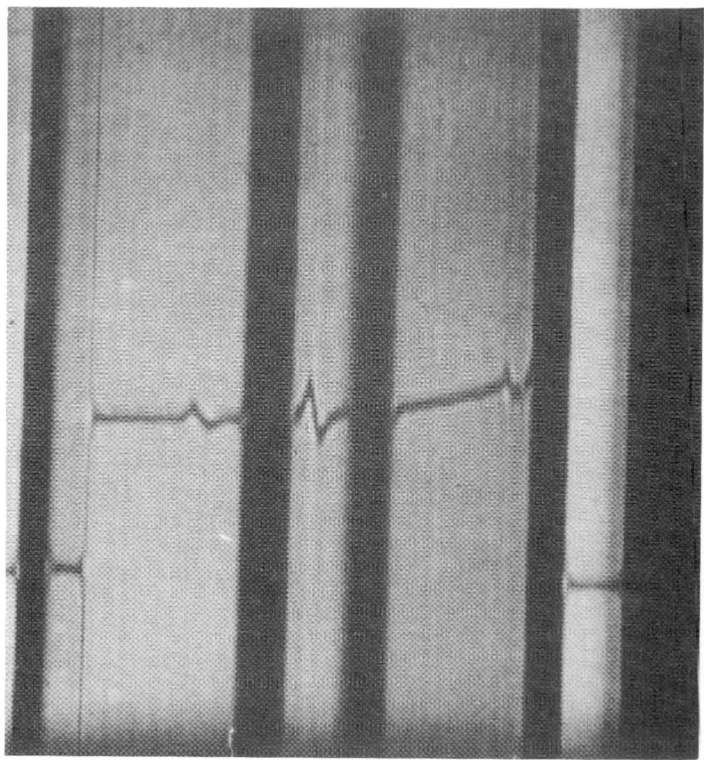

Figure 8. Electron micrograph of the component which condensed between the two partitions. The component was identified as Nβ virus.

The advantage of doing the isopycnic experiments in the analytical centrifuge is that the formation of the gradient and the condensation of the viruses at their specific density levels may be followed with the telescope of the centrifuge and if necessary the gradient may be altered by changing the speed of the rotor. The purpose of the double partition centrifuge cell is to assist in the separation of the individual entities in a multi-component mixture of viruses or high molecular weight protein components. The reorienting gradient rotor was applied to the separation of components of different sedimentation coefficients of the haemolymph of mollusc *Turbo*

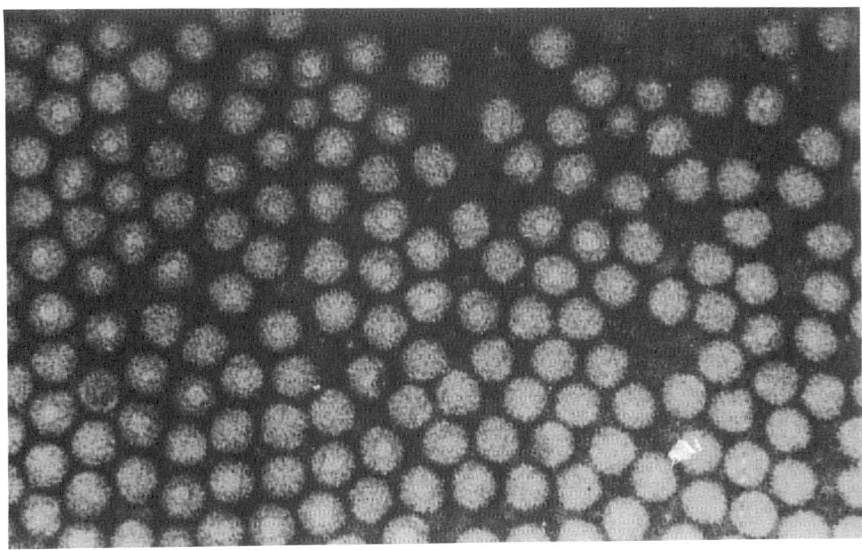

Figure 9. Isopycnic centrifugation diagram in the double partition centerpiece cell. The initial density of the CsCl was 1.340 g/ml; rotor velocity 44700 rpm for 5 hours at 25°C. The component between the partitions, Nγ virus. Note, the tiny components above the upper partition and near the base of the cell.

sarmaticus. It was possible to isolate the two major components of sedimentation coefficients S (108 and 98) as a "single" fraction from the component S44. The S (16-22) was well separated from the S44 component. The separation of S (108 and 98) and S44 was confirmed by electron microscopy.

The reorienting gradient rotor was also applied to the mixture of five viruses obtained from infected and dead caterpillars of the emperor moth *Nudaurelia cytheria cytheria* (N). At best only the main virus Nβ of the mixture could be separated in a pure state. The reason for the limited success may be blamed on the small differences in the sedimentation coefficients of the infective agents and, to a lesser extent, on spreading by diffusion.

Isopycnic centrifugation in a self forming gradient of CsCl enabled the quantitation and isolation of the different N viruses. The isolation in purified states was made possible through the introduction of a separation cell with a centerpiece with two partitions. The two partitions formed an additional chamber two mm wide. During ultracentrifugation of the mixture of the N viruses (α, β, Γ, δ, ϵ) in the self forming CsCl gradient it was essential to select the appropriate initial concentrations of the CsCl. In two centrifugation runs it was possible to isolate and purify N (β, Γ, and ϵ) viruses, the remaining two viruses α and δ could have been recovered separately from the separation cell but they were left on account of their low concentration.

The separation of five morphologically distinct viruses from moribund *Nudaurelia cytheria* caterpillars by isopycnic centrifugation is in agreement with Jucke's[6] results that in purified extracts five antigenically distinct viruses could be isolated by zone electrophoresis in sucrose gradients.

REFERENCES

1. Anderson, N.G., Price, C.A., Fisher, W.D., Canning, R.E., and Burger, C.L. Anal. Biochem. *1*, 1–9 (1964).
2. Beckman Corp. 4-5 Preparative centrifugation.
3. Polson, A. and Kaufmann, K.J. Experientia *26*, 1038 (1970).
4. Polson, A. in Methods in Virology Ed. K. Maramorosch and H. Koprowski. Volume V. Academic Press, New York and London, 1971.
5. Juckes, I.R.M., Russell, B.W. and Polson, A. Preparative Biochemistry *1(2)* 151–161 (1971).
6. Juckes, I.R.M. PhD thesis, University of Cape Town (1972).

Part II - Separation and Purification of Specific Viruses

11. A Rapid Method for the Purification of the MEF_1 Strain of Poliomyelitis Virus*

ABSTRACT

It was shown that on dialyzing infected brain suspensions against buffers of different pH values the virus comes down in the precipitate at pH 4.3 leaving little virus in the supernatant fluid. The virus in the deposit can now be redispersed in a small volume of phosphate buffer of pH 8.2. The substances responsible for the adsorption of the virus to the deposit can be removed by shaking with chloroform without loss of titre of the virus. The virus in a relatively pure state can now be recovered quantitatively by ultracentrifugation.

INTRODUCTION

During attempts at determining the iso-electric point of the MEF_1 strain of poliomyelitis virus in extracts of suckling mouse brain it was noted that the extracts, which were free from coarse particles, became turbid when dialyzed against dilute buffers at pH values below 7. We therefore examined the effect on the titre of the virus suspensions on dialysis against buffers at different pH values. This led to a simple method of purifying

*From: A. Polson and G. Selzer, Bioch et Biophys., *14*:67-70 (1954).

poliomyelitis virus. The fact that SK poliomyelitis virus and encephalomyelitis virus in mice can be precipitated at approximately pH 4.5 and redispersed at pH 8 without loss of activity has been described[1,2].

EXPERIMENTAL

The virus was the MEF_1 strain of poliomyelitis virus adapted to suckling mice in our laboratory$_3$. It was prepared as a 10% suspension of the brains in M/75 phosphate buffer pH 8.2 containing 0.044 M Nacl and 10% rabbit serum. The crude suspension was spun at 2600 rpm for 1 hour and 10 ml portions of the supernatant fluid were dialyzed against 2 litre amounts of phosphate-citrate buffers at different pH values at 4°C for 48 hours. The buffers were those of McIlvaine diluted 1 in 5 with distilled water. The dialysates were spun at 2600 rpm for 1 hour. The deposits were resuspended in M/15 phosphate buffer of pH 8.2, which brought a part only of the precipitate into solution. The resuspended deposit and the supernatant fluid were titrated in albino mice 4 weeks old. Serial tenfold dilutions of the materia! to be tested were made in nutrient broth and each dilution inoculated into groups of 6 mice, each mouse receiving 0.03 ml intracerebrally. The mice were kept under observation for a period of 3 weeks. Titres are expressed as LD_{50} calculated according to the method of Reed and Muench[5].

In a second experiment the brain suspension was mixed with an equal volume of McIlvaine's buffer of twice the normal concentration giving a final pH of 4.3. The solution was not dialyzed but was kept at 4°C for 48 hours and then spun at 2600 rpm for 1 hour. The deposit was resuspended in M/15 phosphate buffer pH 8.2 as in the first experiment and both supernatant fluid and resuspended deposit were titrated in mice as in the first experiment.

In a third experiment an emulsion of 500 infected suckling mice brains in 1000 ml of a 10% serum phosphate buffer M/75 pH 8.2 was centrifuged at 12,000 rpm for 15 minutes. The clear supernatant fluid was separated and the deposit was resuspended in 400 ml phosphate buffer and again centrifuged for 15 minutes at 12,000 rpm. The combined supernatant fluids were dialyzed against 10 litres of McIlvaine's buffer (diluted 1 in 5 with distilled water) at pH 4.3 for 48 hours at 4°C. The dialyzed fluid was spun at 2600 rpm for 1 hour and the deposit resuspended in 200 ml M/15 phosphate buffer of pH 8.2. The suspension was dialyzed a second time as above for a period of 24 hours at pH 4.3, and the deposit after centrifugation, resuspended in 200 ml M/15 phosphate of pH 8.2. This was kept at room temperature for 2 hours, then cooled in iced water and an equal volume of cold chloroform and amyl alcohol mixture (1 part amyl alcohol to 10 parts of chloroform) added[4]. The mixture was shaken in ice-cold water for 15 minutes. After centrifugation at 2000 rpm for 20 minutes the supernatant was treated as before with fresh chloroform-amyl alcohol and again spun at 2000 rpm for 20 minutes.

The supernatant fluid, freed from dissolved chloroform by exposure to a vacuum was centrifuged at 30,000 rpm for 2 hours in the No. 40 rotor of a model L Spinco centrifuge. The pellets were resuspended in 10 ml phosphate buffer of pH 8.2 cooled in ice-water and subjected to chloroform-amyl alcohol treatment for 10 minutes. The chloroform was removed as before and the material centrifuged at 30,000 rpm for 2 hours. To facilitate redispersion of the pellet, 1 ml of concentrated glucose solution was layered at the bottom of the tube before centrifugation. The resulting clear pellet measured 3 to 4 mm in diameter. This was redispersed in 1 ml of 10% rabbit-serum-saline, and the virus content determined by titration in mice.

Because of the small amount of material in the final pellet no direct estimation of its protein content was attempted, but an indirect estimate was obtained from comparison with the size of pellets formed on centrifugation of solutions of haemocyanin which have approximately the same particle size as the poliomyelitis virus. The ml amounts of *Caminella sincta* haemocyanin (sedimentation constant 100 S) of different concentrations were centrifuged at 30,000 rpm for 60 minutes in tubes similar to those used for the final concentrations of MEF_1 virus. The diameters of the pellets were measured. By interpolating the size of the final pellet of the MEF_1 poliomyelitis material into this curve the amount of sedimentable protein present could be estimated.

RESULTS

In Fig. 1 are given the results obtained in two of the first experiments. The interrupted line represents the titres (expressed as negative log LD_{50}) of the supernatant fluid after dialysis against buffer at pH 4.3 while the solid line is that of the deposits resuspended in an amount equivalent to the original volume.

It will be noted that with lowering of the pH the titres of the supernatant fluids become progressively reduced until a pH of approximately 4.3 is reached whereafter the titres again increase. The reverse occurs in the case of the deposit where the maximum titre is reached at approximately pH 4.3.

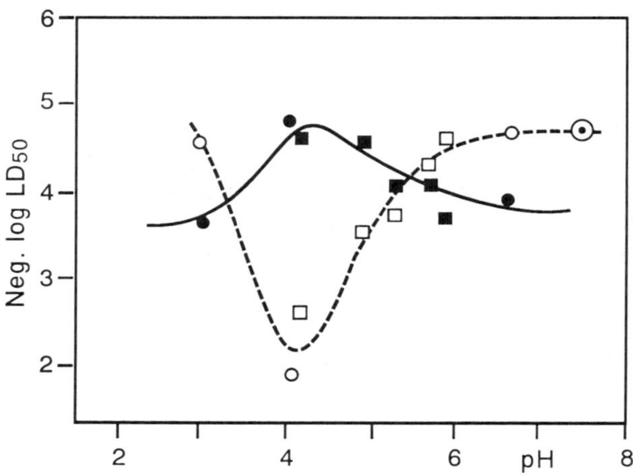

Figure 1. The open circles and squares represent the tires of the supernatants in two experiments respectively (interrupted line). The value at pH 7.5 is the average titre of the original undialyzed emulsion.

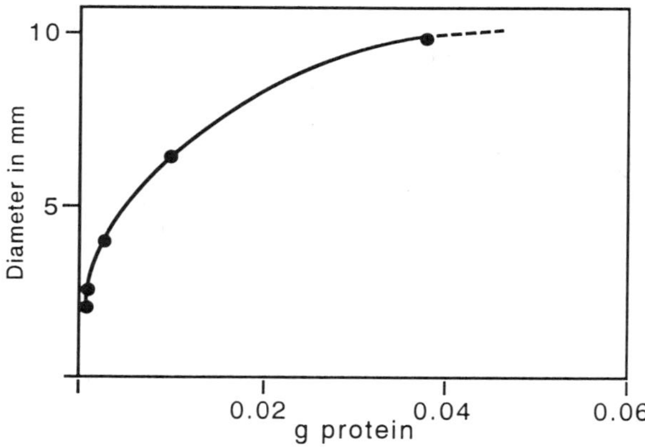

Figure 2. Curve showing the relationship between amount of protein and the diameter of the pellet. The protein spun was the haemocyanin of *Caminella sincta* with sedimentation constant 100 S. It was subjected to centrifugation sufficient to deposit all the protein present.

In the second experiment carried out at higher salt concentration and without dialysis it was observed that the supernatant fluid was not as clear as that obtained against dilute buffer of pH 4.3. Moreover there was considerably more virus in the supernatant fluid ($LD_{50} = 10^{-4.5}$) than in that from the dialyzed material ($LD_{50} = 10^{-2.0}$) indicating that a low electrolyte content is a necessary factor in the removal of the virus from the supernatant fluid.

In the third experiment the LD_{50} of the original suspension (14000 ml) was $10^{-4.5}$ and that of the resuspended pellet (1 ml) $10^{-7.8}$ indicating a quantitative recovery of the virus within the bounds of experimental error.

By the indirect method we have described it could be shown that centrifugation and chloroform extraction has resulted in a significant degree of purification. Thus virus was recovered almost completely from 1400 ml of crude suspension containing 13.7 g of non-dialyzable solid material but in the final pellet containing all the virus there was only 1.1 mg or 0,008% of the total non-dialyzable solid material of the original emulsion. In Fig. 2 the results are recorded graphically.

DISCUSSION

It would appear from these experiments that extracts of the brains of suckling mice infected with MEF_1 strain of poliomyelitis virus contain one or more substances which can be used in the removal of virus from solution. The substances precipitate during dialysis against dilute buffer at pH 4.3 apparently adsorbing the virus from solution. On resuspending the precipitate at pH 8.2 the virus is brought back in suspension while at least one of the brain substances remains undissolved. The virus can be further freed from protein by chloroform treatment without detectable loss its titre, and can be easily concentrated by ultracentrifugation producing a very small pellet. The purity of this final product still needs confirmation by electron microscopy.

It is known that nuclear proteins remain in solution after chloroform treatment. The fact that the poliomyelitis virus does the same may be an indication of its chemical constitution.

ACKNOWLEDGEMENTS

We would like to thank Professor M van den Ende and Dr. T. H. Mead for their keen interest and valuable discussions during the course of our work.

This work was supported by the South African Council for Scientific and Industrial Research. Additional grants for this work have been received from the Nkana-Kitwe and Chingola Poliomyelitis Research Funds.

REFERENCES

1. J. Bourdillon and D. H. Moore, Science, 96, (1942) 541.
2. M. Theiler and S. Gard, J. Exp. Med., 72, (1940) 49.
3. G. Selzer, M. Sacks and M. van den Ende, S.A. Med. J., 26 (1952) 201.
4. M. G. Sevag, D. B. Lackman and J. Smolens, J. Biol. Chem., 124, (1938) 425.
5. L. J. Reed and H. Muench, Am. J. Hyg., 27, (1938) 493.

12. Separation of Small Infective Components of MEF_1 Poliomyelitis and Horsesickness Viruses by Migration into Agar Gels*

ABSTRACT

From gel diffusion experiments it is concluded that the slower sedimenting components of MEF_1 poliomyelitis and African horsesickness viruses are definitely smaller particles than the infective components with higher sedimentation constants. They were capable of migrating into agar gel which held back the components with higher sedimentation constants. The possibility that their differing sedimentations on centrifugation depend on factors other than size, such as lipoid content, can therefore probably be excluded. In the case of horsesickness virus an additional infective component with a sedimentation constant of approximately 100 S was shown. This is additional to the two components with sedimentation constants of 180 S and 476 S which have been previously identified (Polson and Madsen[4]).

INTRODUCTION

Several of the small viruses which have been analyzed in the Spinco preparative ultracentrifuge by the method of Polson and Linder[1] have shown the presence of infective particles of widely different sedimentation

---

*From: A. Polson, Biochim. Biophys. Acta *19*:53-57 (1955), with permission.

constants. Thus the MEF_1 strain of poliomyelitis showed the infectivity to be associated with two components of sedimentation constants 100 S and 170 S (Selzer and Polson[2]). Neurotropic Rift Valley fever which had been passaged intracerebrally for 106 generations in mice followed by 50 passages in chick embryos and a further 9 intracerebral passages in mice showed infective particles of sedimentation constants 175 S and 492 S (Naude, Madsen and Polson[3]) and African horsesickness virus has been shown to have its infectivity associated with particles of sedimentation constants 180 S and 476 S (Polson and Madsen[4]). Neurotropic yellow fever virus on the other hand is an exception in that it has all its infectivity associated with a particle of sedimentation constant 170 S (Polson[5]).

These differences in sedimentation constants may be interpreted in several ways, viz

(a) The particles may be covered with layers of lipoid of different thickness which would thereby give the particles overall lower densities and consequently different sedimentation constants.
(b) The particles may be of different shapes so that they have different frictional constants and consequently different sedimentation rates.
(c) The particles may have the same shape (possibly spherical) but different diameters.

Most of the animal viruses which have been purified and identified on electron micrographs are spherical or nearly spherical and no case has yet been found which showed rod-shaped particles as some plant viruses do, notably tobacco mosaic virus. Possibility (b) is thus very unlikely.

To decide between possibilities (a) and (c), use was made of their abilities to migrate into agar gels of different concentration. It has been found that the migration of proteins (including viruses) into agar gel depends on the concentration of agar in the gel.

METHODS

Standardization of the Agar Gel

By layering a dilute gel over a more concentrated gel an osmotic pressure gradient can be established across the dilute gel. If a watery suspension of protein is placed on the surface of the dilute gel, some fluid will be drawn into the gel, and if the pore sizes in the gel are larger than the protein particles the protein will move into the dilute gel with the water. If pigmented protein is used the movement into the gel can be followed by the movement of the colour band. The osmotic pressure gradient across the dilute gel layer can be varied at will by the incorporation of different concentrations of glycerine in the concentrated gel in the lower part of the tube. In the present experiments this was avoided because the proteins were drawn into higher gel concentrations than when glycerine was omitted. Two haemocyanins of different molecular weight were used in the standardization of the agar gel. That from the crayfish *Jasus lalandii* has a particle diameter of 10.5 mμ and a molecular weight at 450,000 (Joubert[6]); and that from the whelk *Caminella sincta* has sedimentation constant 100 S, a probable molecular weight of 6,600,000 and a particle diameter of 24 mμ (Polson and Linder loc. cit.). In addition to these, rabbit haemoglobin was also used.

A 5 cm long column of 8% agar dissolved n M/15 phosphate buffer of pH 7.2 was allowed to set in each of a series of 7/8" x 6" test tubes. Over these concentrated gel columns a 5 mm thick layer of agar varying in concentration in the different tubes from 0.7% to 8% were placed. A separate set was prepared for each protein tested. A small amount (0.25 ml) of the protein solution under test was placed on the upper surface of the agar in each tube and allowed to remain in contact for three days, the tubes being stored in the cold. The surfaces of the different gels were then washed free of unabsorbed protein and the gels examined for the presence of protein which had penetrated it. It was found that rabbit haemoglobin migrated into the gel to the same extent irrespective of its agar content within the range tested. *Jasus lalandi* haemocyanin moved into those gels which contained less than 7% agar, whereas for the haemocyanin of *Caminella sincta* the limiting concentration was approximately 3%. The relationship between the particle's diameter and the minimum agar concentration which prevents its entrance under the particular conditions of our experiments is probably that shown in Fig. 1. Very dilute agar gels in which the molecules are far apart, would hold back large particles only and very concentrated gels with their molecules close together would exclude small particles. The following equation can be applied:

$$Cd = K \quad (1)$$

In this equation d is the diameter in mμ of the particle and C the limiting concentration of the agar. K is a constant. The value of K calculated from the results obtained with the two haemocyanins is approximately 70.

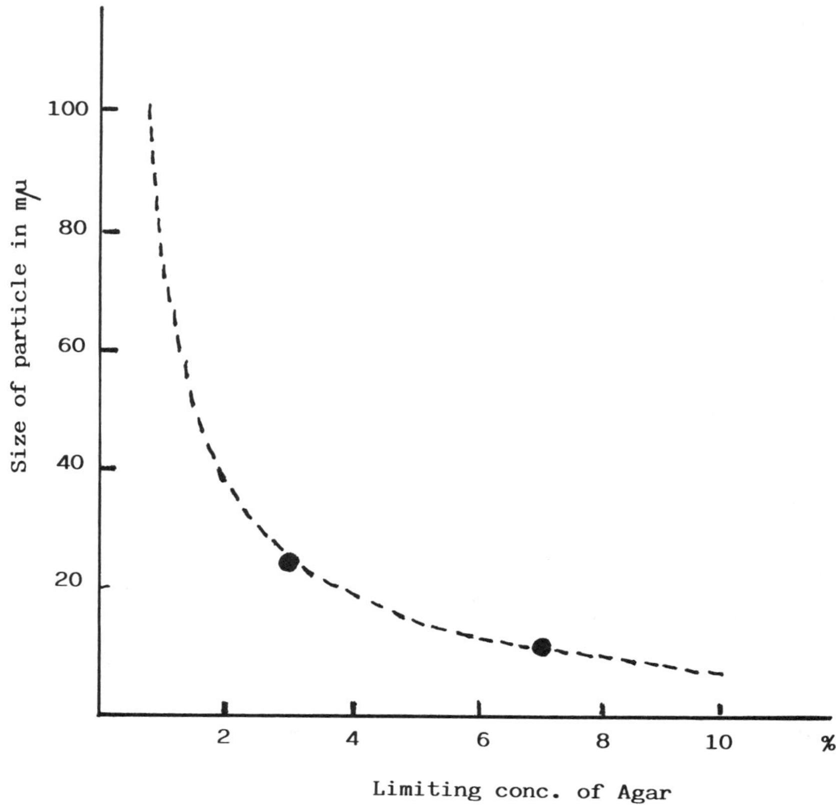

Figure 1. Relationship between size of particle and limiting concentration of agar. (Lowest concentration of agar preventing entrance of particle into it.)

Equation (1) enables one to calculate the approximate limiting concentration of agar for the smaller animal viruses.

RESULTS

MEF_1 Poliomyelitis

From equation (1) it was calculated that the 170 S (30 mμ) particle of MEF_1 virus would be held back by 2.5% agar. To test this the virus from 20 infected suckling mouse brains was extracted in 20 ml 10% rabbit serum

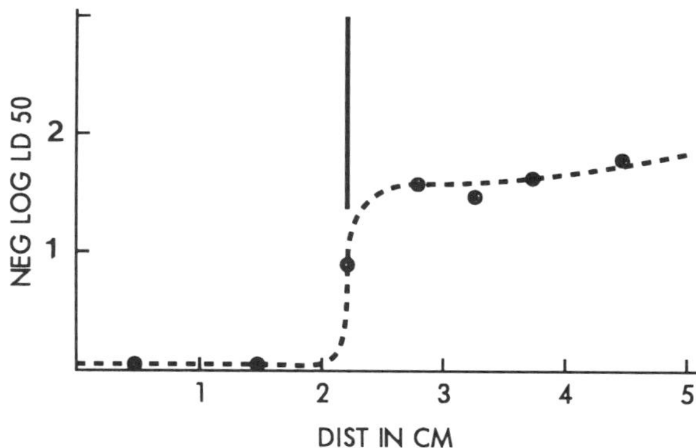

Figure 2. Sedimentation diagram of component with sedimentation constant 100 S isolated from MEF_1 poliomyelitis by migration into 2.5% agar. The solid vertical line represents the position of the haemocyanin (*C. sincta*) boundary. Centrifugation at 20,000 r.p.m. for 100 minutes.

saline and partially purified by differential ultracentrifugation. The final pellet was dispersed in 1 ml 10% rabbit serum saline and 0.25 ml of this suspension was placed in each of 4 tubes prepared with a top layer of 2.5% agar and the usual bottom layer of 8% agar. They were left in the refrigerator for 3 days and during this period nearly all the fluid diffused into the agar. The surface of the agar was thoroughly washed by impinging a light jet of 10% serum saline on to it. After four such washings the tubes were inverted and left to drain. The 2.5% agar layers were then scooped out and ground up in 10 ml serum saline with the aid of a Ten Broeck grinder. The agar was centrifuged off at 10,000 r.p.m. for 10 min and the supernatant fluid mixed with an equal amount of *Caminella sincta* haemocyanin used for particle size determination by the ultracentrifugation method. After centrifugation at 20,000 r.p.m. for 100 minutes in the Model L Spinco preparative ultracentrifuge successive layers were removed and the virus content of each determined by titration in mice. Fig. 2 records the sedimentation diagram of the virus and shows the position or the haemocyanin boundary. There is a single sedimentation boundary corresponding to the virus particle of 100 S. No evidence is found of the 170 S particle normally also present in material not previously subjected to diffusion into 2.5% agar.

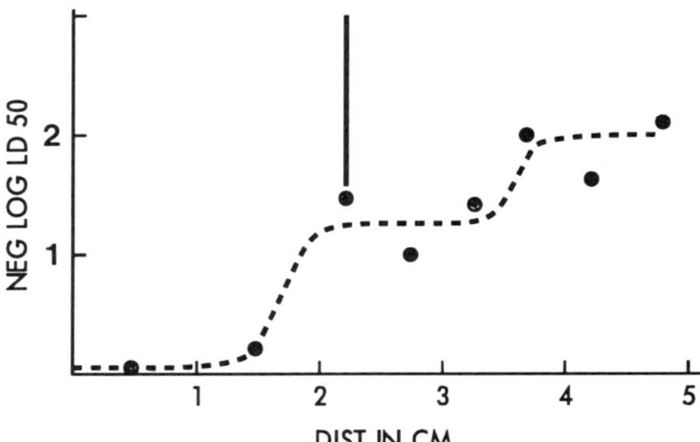

Figure 3. Sedimentation diagram of MEF_1 poliomyelitis virus isolated by migration into 1.5% agar. The solid vertical line represents the position of the haemocyanin (*C. sincta*) boundary. Centrifugation at 20,000 r.p.m. for 100 minutes.

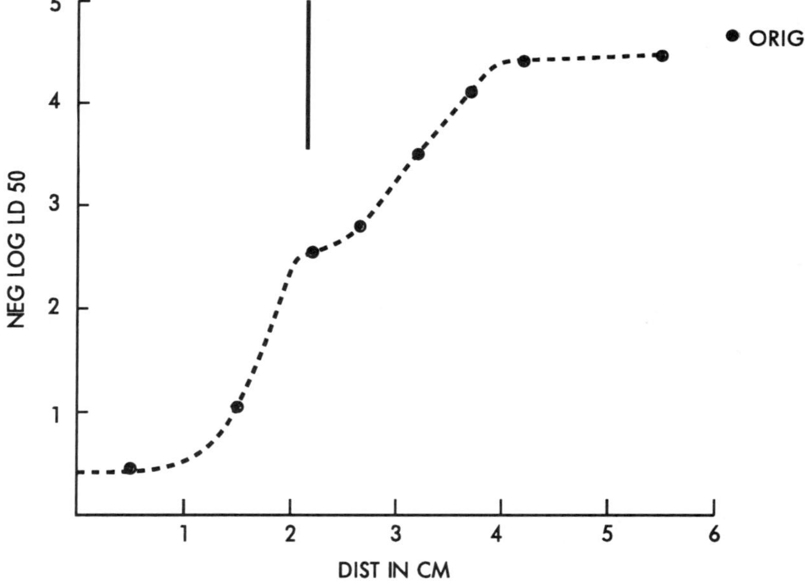

Figure 4. Sedimentation diagram of MEF_1 poliomyelitis virus extracted from infective material ground up in the presence of 2.5% agar. The position of the haemocyanin (*C. sincta*) boundary is indicated by the solid vertical line. Centrifugation at 20,000 r.p.m. for 100 minutes.

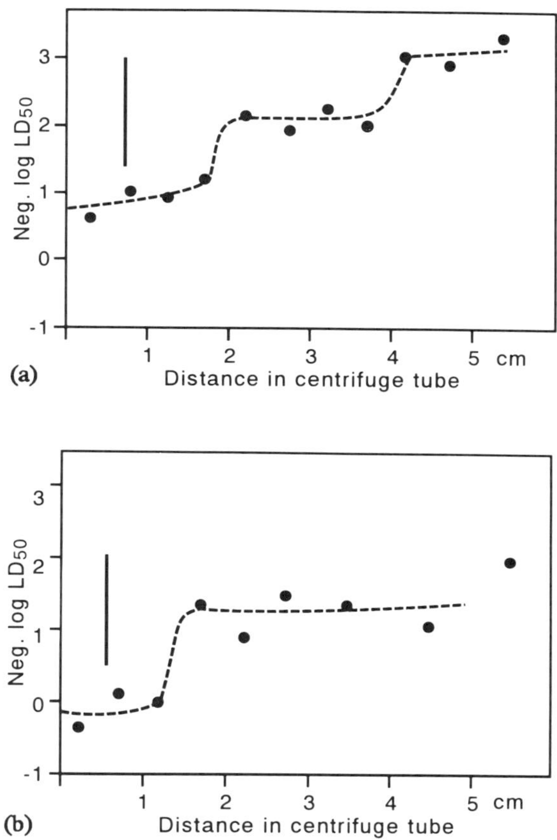

Figure 5. (a) Sedimentation diagram of unfractionated Horsesickness virus. (b) Sedimentation diagram of Horsesickness virus isolated by migration into 1.5% agar. The positions of the haemocyanin (*C. sincta*) boundaries are indicated by the solid vertical lines. Centrifugation at 11,000 r.p.m. for 100 minutes.

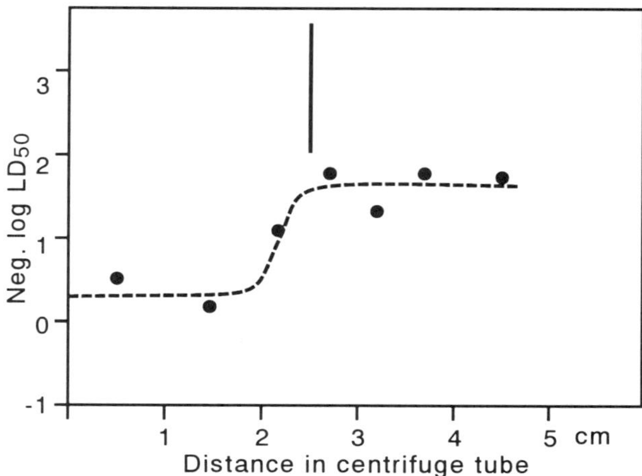

Figure 6. Sedimentation diagram of Horsesickness virus isolated by migration into 1.5% agar gel and ultracentrifuged at 20,000 r.p.m. for 100 minutes. The vertical solid line indicates the position of the reference haemocyanin (*C. sincta*).

Fig. 3 records the results of a similar experiment in which the upper gel layer contained only 1.5% agar. The normal sedimentation pattern indicating the presence of virus of two particle sizes was obtained.

It was shown experimentally that it was not simply the grinding of a virus suspension in the presence of 2.5% agar gel which gave rise to the altered sedimentation diagram. A sedimentation diagram of a virus suspension treated in such a way is recorded in Fig. 4. From these experiments it is concluded that the slower sedimenting component of MEF_1 poliomyelitis virus is a particle of smaller diameter than the component with a sedimentation constant of 170 S and that its slow sedimentation is not due to a higher lipoid content.

Neurotropic African Horsesickness Virus

From equation (1) it can be calculated that the 50 mμ particle of horsesickness virus which has a sedimentation constant of 476 S, would be held back by agar gel with a concentration of approximately 1.5%. The partially purified virus of the strain Vryheid obtained from 20 infected adult mouse brains was suspended in 1 ml 10% rabbit serum saline and allowed to migrate into 1.5% agar gel under similar conditions as was done with the MEF_1 virus. The extract of the 1.5% agar gel was ultracentrifuged at 11,000 r.p.m. for 100 minutes in the presence of *Caminella sincta* haemocyanin (sedimentation constant 100 S). As a control, material not subjected to migration into agar was also ultracentrifuged. Figs. 5 (a) and (b) record the results of a typical experiment. Components of both the main particle sizes are present in the untreated material and only the component with a sedimentation of 180 S is present in the material recovered from the gel. Unlike the sedimentation diagrams obtained with the MEF_1 virus suspensions (Fig. 2 and 3) those obtained with horsesickness virus suspensions show considerable amounts of virus above the two sedimenting boundaries. Attempts to isolate a still smaller component by migration into a gel of higher concentration were not satisfactory on account of the low titre of the apparently smaller component with consequent large losses during recovery from the agar. An additional not previously recognized component was found on ultracentrifuging the material isolated with 1.5% gel at 20,000 r.p.m. for 100 minutes. The results of such an experiment in which *Caminella sincta* haemocyanin was present as reference are shown in Fig. 6. Like the slower sedimenting component of MEF_1 virus this component has a sedimentation constant of approximately 100 S.

ACKNOWLEDGMENTS

The author wishes to express his gratitude to Professor M. van den Ende for his interest in this work, also Dr. G. Selzer for the supply of the MEF_1 virus and Dr. R. A. Alexander for the horsesickness virus. The technical assistance of Miss T. Madsen is gratefully acknowledged.

REFERENCES

1. A. Polson and A.M. Linder, Biochim. Biophys. Acta, *11*, 199 (1953).
2. G. Selzer and A. Polson, Biochim. Biophys. Acta, *15*, 251 (1954).
3. W. Du T. Naude, T.I. Madsen and A. Polson, Nature, *173*, 1051 (1954).
4. A. Polson and T. Madsen, Biochim. Biophys. Acta, *14*, 366 (1954).
5. A. Polson, Proc. Soc. Exptl. Biol. Med., *85*, 613 (1954).
6. F.J. Joubert, Biochim. Biophys. Acta, *14*, 127 (1954).

13. Isolation of the Small Infective Particle in Adapted MEF_1 Poliomyelitis*

ABSTRACT

By means of a combined process of migration into agar gel and ultracentrifugation, the 24 mμ component of MEF_1 poliomyelitis virus adapted to sucklings has been separated from the 30 mμ particles with which it is normally associated, and grown in adult as well as in suckling mouse brains. The progeny of this separated variant in adult and suckling mice is composed of the 24 mμ particles only. This virus is probably a real variant as it "breeds truly" in the host, i.e. the suckling mouse.

INTRODUCTION

The lack of homogeneity of size of infective particles has been reported to be a characteristic of suspensions of a number of viruses. It has been shown by centrifugation in the preparative Spinco ultracentrifuge that in infective mouse brain suspensions of neurotropic African horsesickness virus, particle sizes of 24, 30 and 50 mμ are demonstrable (Polson and Madsen[1]; Polson[2]). In neurotropic Rift Valley Fever virus (R.V.F.), which had had passages in eggs as well as in mouse brains during adaptation, an infective component of particle size 30 mμ occurred together with a 50 mμ

---

*From: A. Polson and G. Selzer, Biochim. Biophys. Acta, *24*: 597-600 (1957). With permission.

component (Naudé, Madsen and Polson[3]). In suckling mouse adapted MEF_1 poliomyelitis virus, particle sizes of 24 and 30 mμ had been found (Selzer and Polson[4]).

In contrast to these findings, a number of viruses which were the carriers of infectivity have been found to be monodisperse in particle diameter. Thus in Yellow Fever (Polson[5]), Semliki Forest and West Nile viruses (Hampton and Polson[6]), infective particles of only 30 mμ have been noticed. Similarly only the 50 mμ particles were demonstrable in pantropic R.V.F. virus and neurotropic R.V.F. virus. These had received passages in mice only.

Attempts to cultivate the smaller horsesickness particles and maintain them in passage, free from the 50 mμ units, have been unsuccessful. After nine ultracentrifugal separations and limiting dilution passages in mice,the virus still showed the presence of the 50 mμ infective particles (Polson and Madsen[1]). Different results have, however, been obtained with the small infective particle of adapted MEF_1. In this case we have succeeded in isolating the 24 mμ form and showed that its progeny, when inoculated intracerebrally into adult as well as suckling mice is composed of the 24 mμ particles only.

EXPERIMENTAL

The virus purified from 60 infective suckling mouse brains, by isoelectric precipitation, ultracentrifugation, and chloroform treatment, according to the method of Polson and Selzer[7], was suspended in 1 ml of a solution of 10% normal rabbit serum in M/15 phosphate buffer of pH 7.0. This was followed by the isolation of the 24 mμ particle using the agar gel technique of Polson[2]. The agar gel extract which contained the greater portion of the 24 mμ particles was subjected to analysis for particle size using the Spinco ultracentrifuge. The sedimentation diagram obtained is shown in Fig. 1. This diagram should be compared with that in Fig. 2 which is a typical sedimentation diagram of the suckling mouse adapted MEF_1 virus as obtained from infected suckling mouse brain extracts. Although larger-sized virus particles may still be present in the virus extracted from the agar, the proportion of the smaller component is much greater, indicating that the bulk of the 30 mμ component had been removed. To further reduce the amount of the larger component in the material used for passage, a portion of the sample taken between the 2 and 2.5 cm levels (see Fig. 1) was used as inoculum for adult mice. When paralytic symptoms

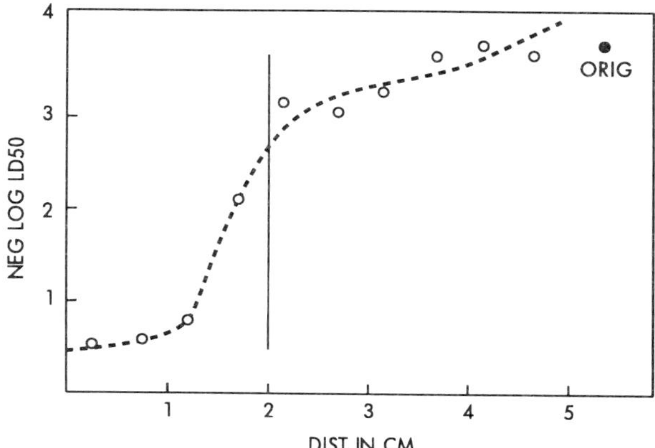

Figure 1. Sedimentation diagram of the smaller component of MEF_1 virus separated by migration into agar gel[1]. Centrifugation of 20,000 rmp for 100 min. The vertical line marks the position of the haemocyanin boundary.

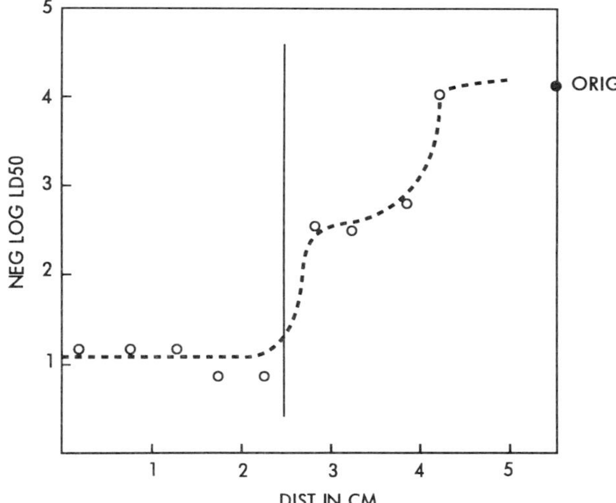

Figure 2. Sedimentation diagram of MEF_1 virus in suckling mouse brain without preliminary selection of the smaller component. Centrifugation at 20,000 rpm for 100 min. The vertical line marks the position of the haemocyanin boundary.

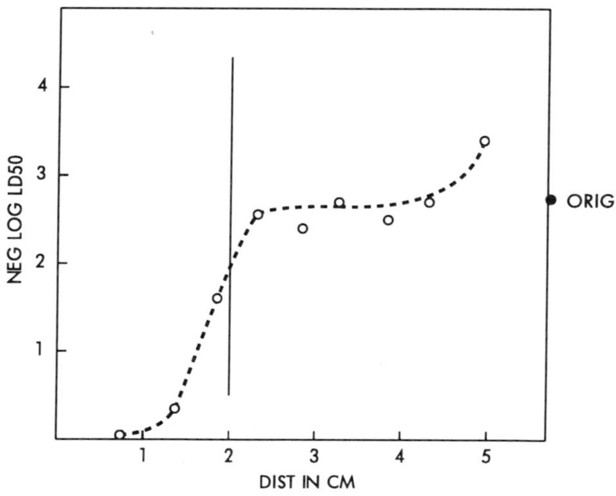

Figure 3. Sedimentation diagram of the 24 mμ component MEF_1 grown in adult mouse brains. Centrifugation at 20,000 rpm for 100 min. The vertical line marks the position of the haemocyanin boundary.

developed, the mice were killed, their brains collected and an emulsion prepared for ultracentrifugation in the presence of *Caminella sincta* haemocyanin. As before, the sample between 2 and 2.5 cm was collected for the next passage in adult mice. This process of selection of the lighter particle was repeated three times. Brains from mice of the third passage were emulsified, and the size of the virus particle in the emulsion determined by centrifugation in the usual way. The result of the centrifugation is shown in Fig. 3. Here only the component of sedimentation constant 100 S (24 mμ diameter) could be seen. That Fig. 3 represents a typical result with a suspension of monodisperse particles is substantiated by the results obtained with *C. sincta* haemocyanin in control experiments. The concentration of haemocyanin at various levels in a centrifuge tube which had been subjected to appropriate centrifugation, as determined by refractometric measurements, is shown in Fig. 4. It will be noticed that the distribution of the virus at various levels was very similar to that obtained with the haemocyanin, thus lending support to the conclusion that the virus was

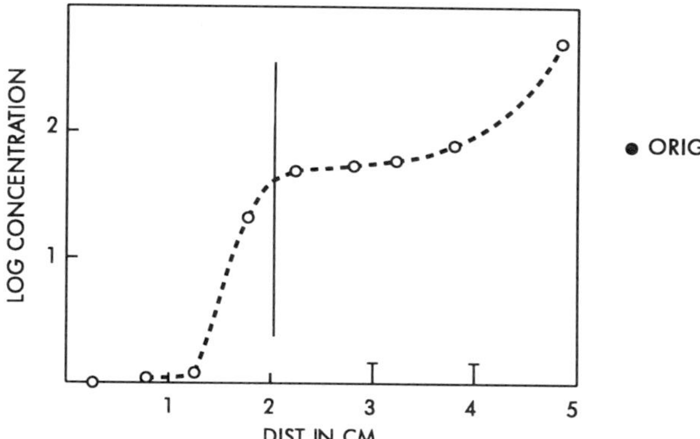

Figure 4. Sedimentation diagram of *C. sincta* haemocyanin. Centrifugation at 20,000 rpm for 100 min. The concentration in the different layers in the tube was plotted exponentially against distance in the tube.

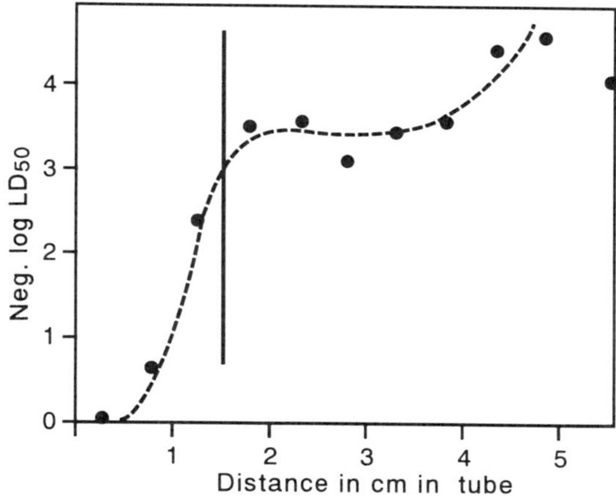

Figure 5. Sedimentation diagram of 24 mμ MEF_1 virus grown for two generations in suckling mouse brains. Centrifugation at 21,000 rpm for 60 min. The vertical line marks the position of the haemocyanin boundary.

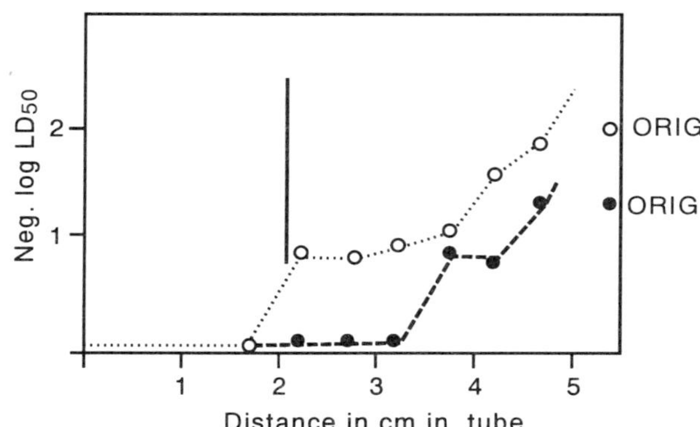

Figure 6. Sedimentation diagrams of the MEF_1 virus before adaptation to suckling mice. In both diagrams there is evidence for the presence of the 24 mμ form of the virus. The vertical line marks the position of the haemocyanin boundary.

homogeneous with respect to particle size. The isolation of the 24 mμ particle in the manner described has been accomplished on three separate occasions.

In Fig. 5 is given a sedimentation diagram of the virus of particle diameter 24 mμ after it had received two further passages in suckling mice. It is interesting to note that the 30 mμ form is still absent and that the titre of the unspun material is now appreciably higher than that of the material (24 mμ strain) which was grown in adult mice (see Fig. 3). The titre is also higher than that which the 24 mμ fraction had when it was present together with the 30 mμ virus in the original unfractionated suckling mouse adapted virus (Fig. 2).

The adaption of the MEF_1 virus to suckling mice is not responsible for he appearance of the 24 mμ particles, since the two forms are present in the original unadapted strain. The sedimentation diagrams depicted in Fig. 6 show the result of two experiments performed on this material.

It can further be mentioned that no consistent results were obtained in experiments with suckling mouse adapted virus which received *further passages in adult mice*! The results obtained from two ultracentrifugation experiments, performed on fresh material on two different occasions, indicated that in the one experiment the 24 mμ virus was predominant and in the other experiment the 30 mμ form was present in highest titre. More experiments along these lines must be done to explain the phenomenon.

At the moment it is not known whether the 24 mμ particle will retain its homogeneity on prolonged passage in adult or suckling mice or in tissue culture, but the object of this report is to supply evidence that the small particle in adapted MEF_1 poliomyelitis is capable of being cultured independently of the larger-sized virus particle.

ACKNOWLEDGEMENTS

The authors wish to express their gratitude to Professor M. van den Ende for his interest in this work and for many helpful discussions. We are grateful to Miss D. Deeks and Mr. T. Norcott for valuable technical assistance.

Additional funds were obtained from the Nkana-Kitwe and Chingola Poliomyelitis Research Funds.

REFERENCES

1. A. Polson and T. I. Madsen, Biochim. Biophys. Acta, *14*, 366 (1954).
2. A. Polson, Biochim. Biophys. Acta, *19*, 53 (1956).
3. W. du T. Naudé, T. I. Madsen and A. Polson, Nature, *173*, 1051 (1954).
4. G. Selzer and A. Polson, Biochim. Biophys. Acta, *15*, 251 (1954).
5. A. Polson, Proc. Soc. Exptl. Biol. Med., *85*, 613 (1954).
6. J. W. F. Hampton and A. Polson. Unpublished results.
7. A. Polson and G. Selzer, Biochim. Biophys. Acta, *14*, 67 (1954).

14. Studies on Poliomyelitis Virus — Concentration and Purification of the Virus*

ABSTRACT

For the purpose of determining the immunogenic potency of polio virus, relatively large amounts of concentrated virus material were prepared which had titres of the order of 10^{10} T.C.I.D.$_{50}$ per ml. These were obtained by pervaporating large quantities of tissue culture fluid containing approximately $10^{6.5}$ T.C.I.D.$_{50}$ per ml.

INTRODUCTION

The perfection of tissue culture methods has led to the development of a formalin-treated vaccine against poliomyelitis which has been widely used. The method generally employed for the preparation of this vaccine involves the treatment of tissue culture fluids, containing not less than 10^6 tissue culture infective doses (T.C.I.D.$_{50}$) of virus per ml. with formalin, at a concentration of 1/1000 to 1/4000.

The unfortunate occurrence of cases of poliomyelitis which could be attributed to residual live virus in one or two batches of vaccine led to the introduction of very strict regulations governing the method of manufacture and of carefully prescribed tests for safety and potency. It is apparent that under the conditions of manufacture at present generally employed, a critical

*From: A. Polson and J. W. F. Hampton, J. Hyg. (Camb.), *55*:334-346 (1957), with permission.

concentration of formaldehyde is required to render the vaccine safe for human administration and at the same time to retain its immunizing power.

The bulk of antigen present per unit volume in a vaccine prepared from tissue culture fluid containing 10^6 T.C.I.D.$_{50}$ per ml. of virus only 30 mμ in diameter is extremely small. Facilities available to us made it possible to prepare virus concentrates containing 1000–10,000 times as much virus per ml. as the vaccine so far used. With such concentrates we have been able to undertake experiments to determine whether less critical, but more reliable, methods could be employed for the inactivation of virus without rendering it antigenically inert.

This paper describes the methods employed in the preparation of concentrated virus suspensions.

MATERIALS AND METHODS

Tissue culture fluids containing the Brunhilde strain (Type 1), Collins (Type 2), and Leon (Type 3), which had been rejected for use in vaccine were used in these experiments. These were produced by the conventional methods involving virus growth on trypsinized monkey kidney cells in culture medium '199'.

Cellophane sausage-casings (The Visking Corporation, Chicago) of about 2 in. diameter when inflated, were fitted with rubber bungs at one end into which a rubber-capped glass tube was inserted for convenience of filling and emptying under sterile conditions. It is advantageous to offset the position of the filling tube close to one side of the bung for the removal of air bubbles and for the complete emptying of the casing. The filling tube should be long enough to enable its cap to be dipped into boiling water before and after filling or emptying; the other end of the casing is sealed in the usual way. The use of long lengths of casings, holding up to 10 liters, is to be recommended as this minimizes the number of filling operations. After sterilization by steaming, these casings were filled by siphoning directly from the 10 liter bottles in which the material was contained. When filled and capped the casings were placed on a grid of 1/2 in. steel mesh having a number of table fans playing on its surface and placed about 3 ft. above a number of domestic 1kW. heaters fitted with reflectors. In these circumstances the rate of evaporation was sufficient to keep the temperature of the fluids at 23°C. or less, where the relative humidity was 50–55%. By this means a concentration to 1/10 vol. can be obtained in about 9 hr. Such a period of evaporation was followed by dialysis overnight at 5° against a volume of distilled water equal to the original volume of material taken.

The casings were then either refilled, or their contents transferred into a smaller number of casings, and the pervaporation and dialysis cycle repeated until a volume of about 200 ml. was reached. At this stage cell debris and other insoluble material was centrifuged out at 2000 r.p.m. for 45 min. Pervaporation was continued until 1/1000th of the original volume was obtained. This was dialysed finally against M/15 phosphate buffer of pH 7.0, centrifuged at 10,000 r.p.m. for 10 min., ampouled in air, shell-frozen and stored in a dry ice cabinet at about -40°C.

RESULTS

A pilot batch of 5 l. of Type 1 virus having an initial titre of $10^{-6.3}$ T.C.I.D.$_{50}$ was processed in this manner to yield 50 ml. of concentrate having a titre of $10^{-8.9}$ T.C.I.D.$_{50}$. This was further concentrated by ultracentrifugation at 30,000 r.p.m. for 90 min. with resuspension of the pellets in 5 ml. phosphate buffer at pH 7.0 to give a colourless, opalescent fluid representing a thousand-fold concentration and having a titre of $10^{-10.6}$ T.C.I.D.$_{50}$ per ml. Taking into consideration the volume of the pellet, it can be assumed that the concentration of active virus in the pellets was in excess of 10^{12} infective units per ml. This showed quantitative recovery of live virus (within the limits of experimental errors). Concentration of larger quantities of virus suspensions was therefore carried out. In this way 62.1 l. of Type 1, 55 l. of Type 2, and 49.7 l. of Type 3 virus were concentrated to 76, 60 and 70 ml. respectively, in 4 to 5 cyc. of pervaporation, and without ultracentrifugation. Penicillin and streptomycin were added in each cycle. During the final clarification at 10,000 r.p.m. a considerable amount of fatty material separated in the supernatant. As the removal of this had been found to give enhanced titres, a sample of each concentrate was extracted by shaking once with an equal volume of ether in the cold before assay.

These concentrated virus suspensions (subsequently referred to as 'crude concentrates') were dark viscous fluids of protein concentration approximately 4.44% (by refractometer). The concentrations of virus (by tissue culture assay) were: Type 1, $10^{10.0}$; Type 2, $10^{9.8}$ and Type 3, $10^{9.5}$ tissue culture infective doses per ml.

After purification for electron microscopy[1,2] as previously described these concentrates gave photographs typical of pure virus suspensions. It is intended to describe their particle sizes and some other properties at a later date.

SUMMARY

For the purpose of determining the immunogenic potency of polio virus, relatively large amounts of concentrated virus material were prepared which had titres of the order of 10^{10} T.C.I.D.$_{50}$ per ml. These were obtained by pervaporating large quantities of tissue culture fluid containing approximately $10^{6.5}$ T.C.I.D.$_{50}$ per ml.

ACKNOWLEDGMENTS

We wish to acknowledge the valuable assistance received from Dr. P.D. Winter and his staff, of the Poliomyelitis Research Foundation, in carrying out the tissue culture titrations of these materials.

REFERENCES

1. Polson, A. and Selzer G. (1954). Biochim. Biophys. Acta, *14*, 67.
2. Hampton, J. W. F., Polson, A. & Selzer, Golda (1955). Biochim. Biophys. Acta, *17*, 592.

15. Investigations of Poliomyelitis Antigens*

ABSTRACT

The complex nature of the antigens in suckling mouse-adapted MEF_1 virus was indicated by Selzer and Polson[1], who found that the infectivity was associated with two components of approximate sedimentation coefficients 100 and 173 S respectively. In addition to these components a "soluble antigen" was found which had a sedimentation coefficient of about 22 S. These results, which were obtained by using a Spinco preparative centrifuge (Polson and Linder[2]), have now been confirmed in the Spinco analytical ultracentrifuge of the Nobel Institute, Stockholm, and other types of poliomyelitis virus have been studied in both instruments.

MATERIALS AND METHODS

Poliomyelitis Virus

The strains of virus used in this study were those incorporated in the Swedish poliomyelitis vaccine. These are:

Type I strain 1423/53, here referred to as Stockholm Type I,
Type II MEF_1,
Type III Saukett and
the Cincinnati strain of Type I was also used in some experiments.

---

*From: A. Polson, A. Ehrenberg and R. Cramer, Biochim. Biophys. Acta *29*: 612-622 (1958). With permission.

Purification

The method of Polson and Hampton[3] was followed. Infective tissue-culture fluid (5-20 liters) was concentrated by pervaporation (drying causes inactivation of the virus and must be avoided during the operation) to about one-fifteenth of its original volume, dialysed against distilled water overnight and deproteinized by being briskly shaken with two volumes of chilled chloroform at 0°. The mixture was centrifuged cold and the clear aqueous layer treated with chloroform as before. After centrifugation the aqueous layer was extracted with light petroleum in order to remove dissolved chloroform and further concentrated by pervaporation to about one-hundredth of its original volume. The concentrate was dialyzed overnight against saline and purified by differential centrifugation. Three cycles of centrifugation were found sufficient for this purpose. The first was done at 30,000 r.p.m. for 90 min to bring down the virus and heavy particles which had not been removed by the chloroform. The second was done on the resuspended pellet material to remove coagulated particles. This was done at 12,000 r.p.m. for 10 min. The third cycle of centrifugation was again done at 30,000 r.p.m. for 90 min to bring down the virus in its final purified form. The purity was assessed by electron microscopy. Figs. 1, 2 and 3 show typical photographs.

Helix Pomatia Haemocyanin

Blood from the hibernating animals was centrifuged at 10,000 r.p.m. for 15 min. The pigment was sedimented by centrifugation at 30,000 r.p.m. for 90 min and re-suspended in M/15 phosphate buffer at pH 6.8. The pigment was used within 3 h after the collection of the blood.

Measurement of Sedimentation Coefficients in the Preparative Spinco Centrifuge

Crude tissue culture fluid, after preliminary clarification at 10,000 r.p.m. for 15 min, was centrifuged in the presence of 0.5% *Helix pomatia* or *Caminella* (now *Burnupens*) *sincta* haemocyanin, which provided references for the calculation of sedimentation coefficients and a density gradient in which convection-free sedimentation took place. After centrifugation, the position of the haemocyanin boundary was recorded and the contents of the tube sampled at 0.5 cm intervals with the precautions described by Polson

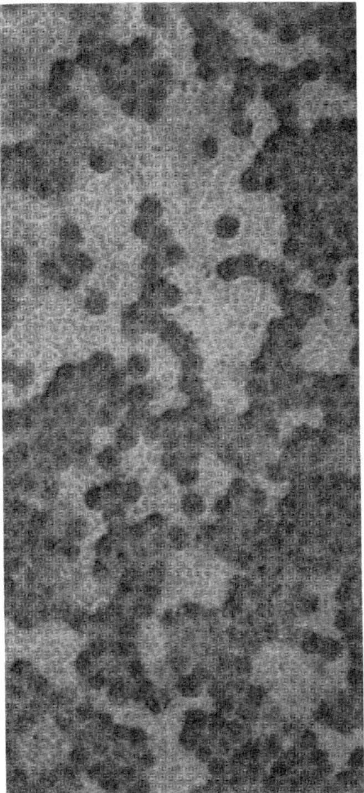

Figure 1. Electron micrograph of Type III virus, Saukett strain. Chromium shadowed at 90°. x 88,000.

and Madsen[4]. The position of the virus boundary was determined and the sedimentation coefficient calculated from the relationship

$$S_1 = S_2 \frac{H_1}{H_2} \frac{(2X + H_2 \text{ Sin } \alpha)}{(2X + H_1 \text{ Sin } \alpha)}$$

In this equation, S_2 is the sedimentation coefficient of the haemocyanin, H_2 is the distance through which the haemocyanin boundary moved from the meniscus, S_1 and H_1 are the sedimentation coefficient of the virus and the distance through which its boundary sedimented respectively, X is the

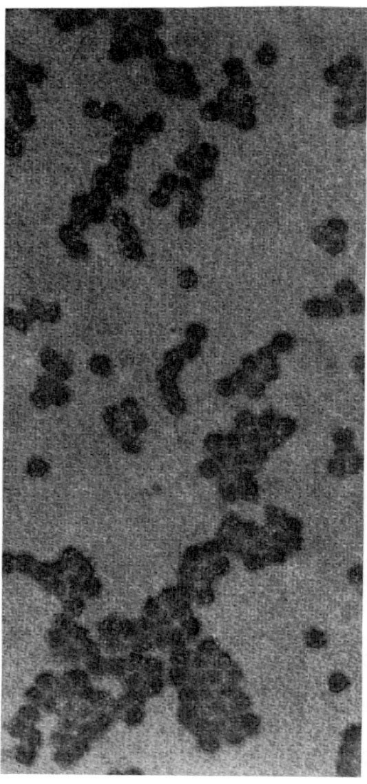

Figure 2. Electron micrograph of Type II virus, MEF_1 strain. Chromium shadowed at 90°. x 88,000.

distance of the meniscus from the centre of rotation, in the present experiments 5.34 cm, and α is the angle of inclination of the tube with the rotor axis during centrifugation, and equal to 26°.

The Sedimentation Coefficients of Haemocyanins

As the haemocyanins of Caminella sincta and Helix pomatia are important as reference substances for measuring the particle sizes in the preparative centrifuge, their sedimentation coefficients were re-determined in the Spinco analytical centrifuge. Discrepancies recently noticed by Miller and Golder[5] between the results given by this instrument and by the oil

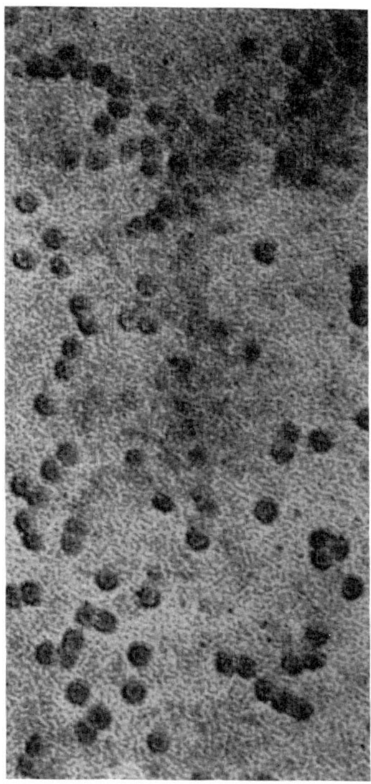

Figure 3. Electron micrograph of Type I virus, Stockholm strain. Chromium shadowed at 90°. x 88,000. Note the considerable amount of relatively low molecular material in the background.

turbine centrifuge (used for earlier work on haemocyanins) made re-determination essential.

The measurements were made in M/15 phosphate buffer pH 6.8 using purified haemocyanin at a concentration (0.5%) close to that employed in experiments with the preparative centrifuge. The results (Table 1) show that both haemocyanins were inhomogeneous in spite of being fresh when examined and at a pH in the middle of their pH-stability ranges. Sedimentation diagrams are given in Figs. 4 and 5. The two peaks in the *C. sincta* diagram separated gradually and are therefore not artifacts caused by convection.

Table 1. Sedimentation Coefficients of *H. pomatia* and *C. sincta* Haemocyanins

0.5%, in M/15 phosphate buffer of pH 6.8

Materials	Sedimentation coefficients (% of total) Svedberg units
H. pomatia	87 (80%), 63 (20%)
C. sincta	93.3 ± 2.0, 88.9 ± 1.2; average 90 ±3

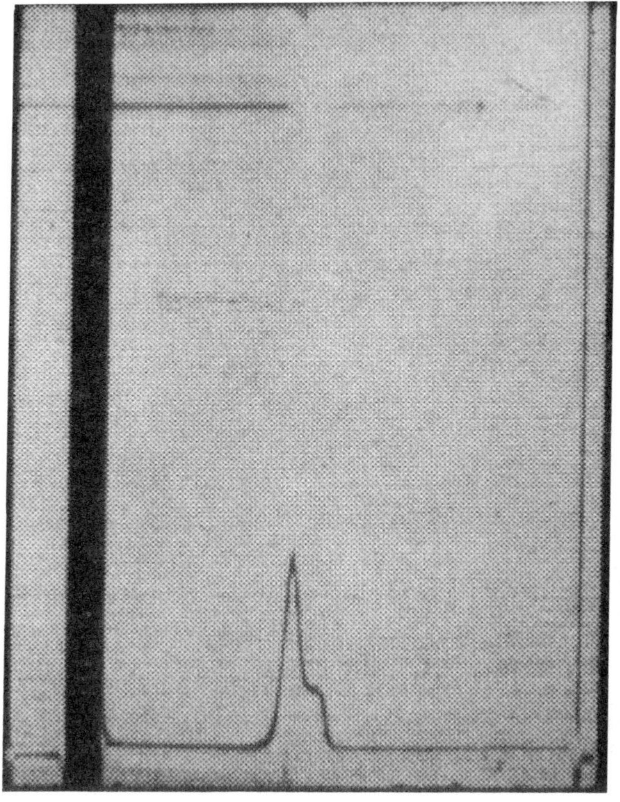

Figure 4. Sedimentation diagram of 0.5% *Helix pomatia* haemocyanin in *M*/15 phosphate buffer at pH 6.8. Exposure taken 39 min after boundary formation in the synthetic boundary cell and 36 min after attaining the equilibrium speed, 14,290 rpm.

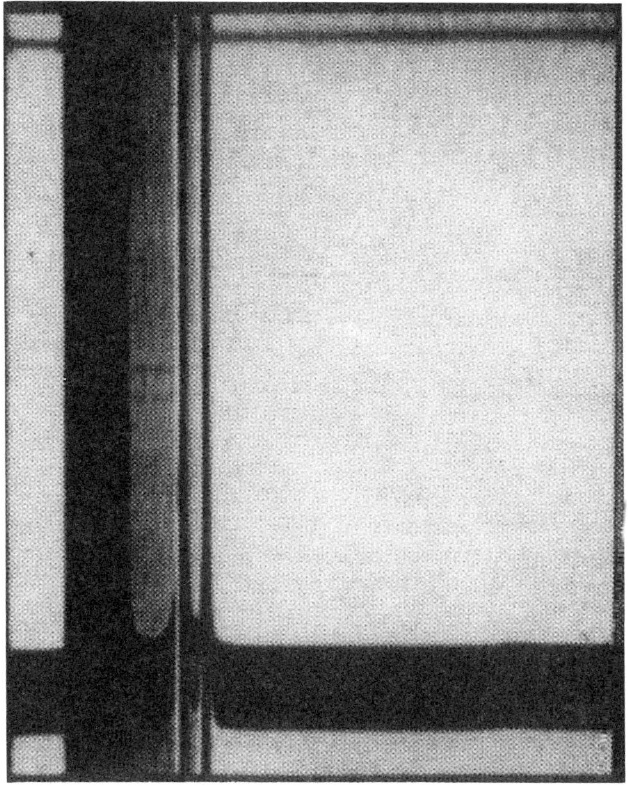

Figure 5. Sedimentation diagram of *Caminella sincta* haemocyanin in *M*/15 phosphate buffer at pH 6.8. Exposure taken after 48 min at 29,500 rpm. In this experiment an ordinary cell and the inclined bar assembly was used.

The haemocyanin of *H. pomatia* was reported to be homogeneous by Svedberg and Pedersen[6] and by Brohult[7]. In these cases the haemocyanin was subjected to ultracentrifugation without any preliminary purification. In the present work, however, the material was freed from low molecular and mucoid material by one cycle of ultracentrifugation. Whether this procedure caused a portion of the molecules to dissociate into smaller units cannot be stated with certainty.

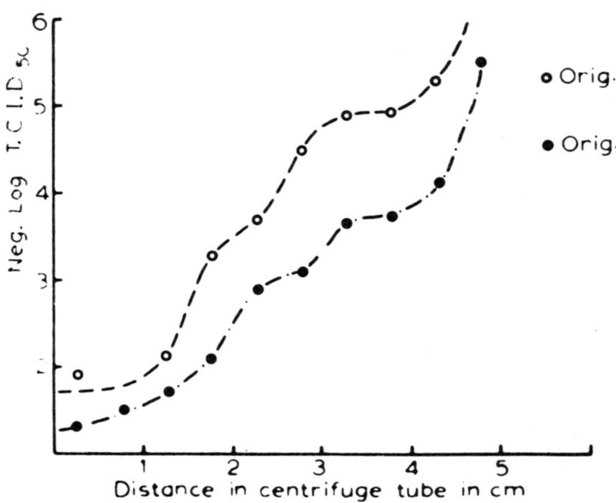

Figure 6. Sedimentation diagrams of Type I virus, Cincinnati strain. Both diagrams were obtained by centrifugation at 20,000 rpm for 100 min in the preparative rotor of the model L Spinco centrifuge.

Measurement of Sedimentation Coefficients in the Spinco Analytical Ultracentrifuge

The centrifuge was a model E Spinco equipped with a phase plate and a rotor temperature controller. The synthetic boundary cell was used because of its large opening, which lessened the possibility of droplet formation when the cell was filled and emptied. Between experiments the cell was sterilized with formalin followed by thorough washing with water and finally with alcohol. All runs were made in 1% sodium chloride. The rotor velocity was 14,290 r.p.m. in most experiments and its temperature was kept close to 20°. In a few cases the speed was increased near the end of the run to improve the resolution of the slower sedimenting components.

The sedimentation coefficients were determined in the usual way and corrected to standard conditions. In a few favorable cases attempts were made to calculate diffusion constants by the methods of Ehrenberg[8].

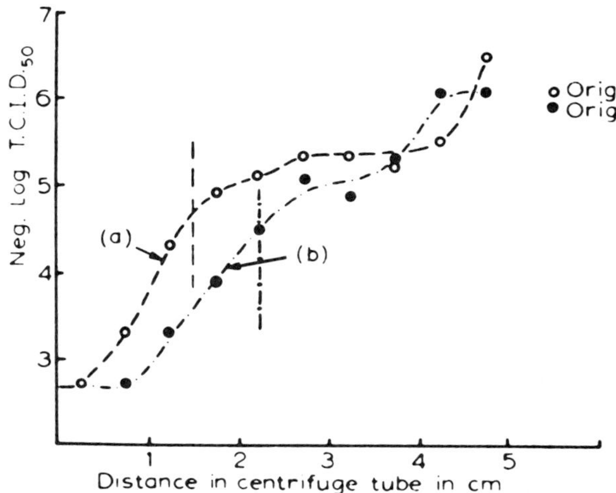

Figure 7. Sedimentation diagrams of Type I virus, Stockholm strain. (a) was obtained by centrifugation at 21,000 rpm for 60 min and (b) at 20,000 rpm for 100 min. The vertical lines indicate the boundaries of the faster sedimenting *H. pomatia* haemocyanin component.

RESULTS

Ultracentrifugation in the Preparative Rotor

The Cincinnati and Stockholm strains of Type 1 and the MEF_1 strain of Type II were first examined in order to find a strain having enough of the slower sedimenting component to be readily detected by light-refraction methods in the analytical ultracentrifuge.

In Figs. 6, 7, 8 and 9 are given sedimentation diagrams of Cincinnati, Stockholm and MEF_1 viruses, respectively. The ordinates indicate the titers and the abscissae the distances from the menisci at which the samples were taken. The data on the Stockholm virus are from three experiments performed under different conditions of centrifugation. The vertical lines indicate the positions of the haemocyanin boundaries at the end of the ultracentrifugation run.

In the diagrams obtained with the Cincinnati and MEF_1 strains there are two distinct "plateaus", which indicate that the virus infectivity was

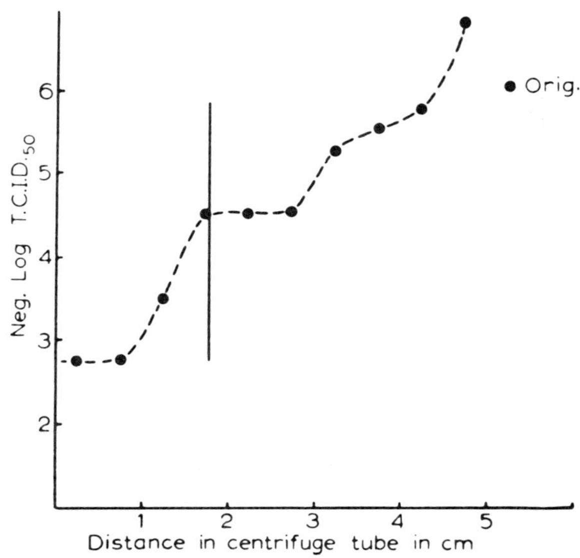

Figure 8. Sedimentation diagram of Type I virus, Stockholm strain, obtained by centrifugation at 18,500 rpm for 90 min. The vertical line indicates the average position of the boundaries of the *C. sincta* haemocyanin with which the virus was mixed prior to ultracentrifugation.

associated with two components having different sedimentation coefficients. The activity of the slower sedimenting component comprises approximately 3-10% of the total. In the experiments with the Stockholm strain no distinct second sedimenting boundary could be detected with certainty in (a), and this was interpreted as indicating that the two sedimenting components were present in about equal amounts. In Fig. 7(b) and 8, however, two distinct boundaries are visible. It would appear that the slower sedimenting component comprised 10-30% of the total infectivity. Further details are given in Table 2.

In the experiment (a), Fig. 7 on the Stockholm strain, the faster sedimenting boundary was assumed to be at 2.5 cm, at the position of the slight rise in the diagram. The "50%" boundaries were calculated by the method Polson and Selzer[9].

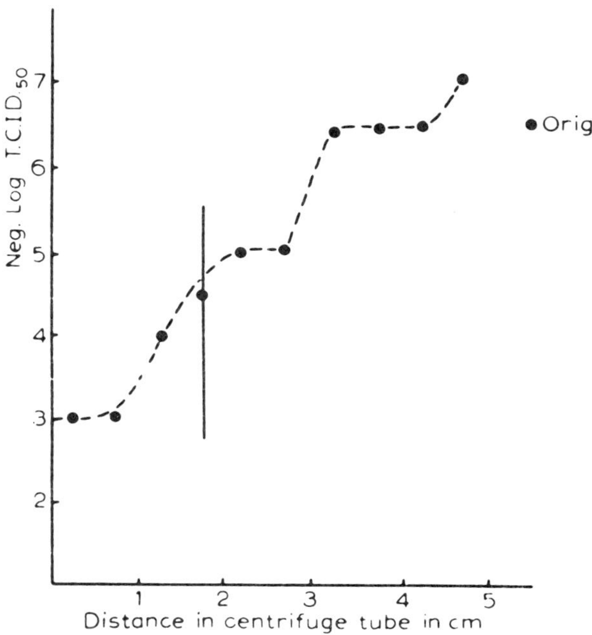

Figure 9. Sedimentation diagram of Type II virus, MEF$_1$ strain, obtained at 20,000 rpm for 100 min. The vertical line indicates the average position of the boundaries of the *C. sincta* haemocyanin components.

Table 2. Ultracentrifugation in the Preparative Rotor
Reference substances 0.5% *Helix pomatia* (H.p.) and
Caminella sincta (C.s.) haemocyanins
a = 26° and X = 5.34 cm

Virus Strain	Reference Substance	Sedimentation Coefficient I	II	Approximate % of infectivity I	II
Cincinnati	None			97	3
				90	10
Stockholm	(a) H.p.	150	86	70	30
	(b) H.p.	147	86	90	10
	(c) C.s.	156	90	90	10
MEF$_1$	C.s	155	90	97	3

Table 3. Results of Sedimentation Analysis of Polio Virus with the Spinco Analytical Ultracentrifuge

Experiment No	Material	Total concentration* mg/ml	$S°_{20}$ in Svedberg units (relative concentration) in % ** I	II	III	$D°_{20}$ in Fick units for fastest component
1	MEF_1 Cape Town	0.21	157 (90-99%)	90 (trace)	-	
2	MEF_1	0.87	156 (90-99%)	80 (trace)	20 (trace)	1.4
3	MEF_1	1.16	156 (79%)	73 (8%)	25 (13%)	1.7
4	MEF_1	1.6	157 (70%)	70 (30%)	20	
5	Stockholm I	0.23	150 (trace)	94 (90-99%)	-	
6	Stockholm I	3.4	155 (42%)	85 (31%)	20 (27%)	1.5
7	Stockholm I Sep. Cell	3.0	159 (45%)	81 (27%)	16 (28%)	
8	Saukett	0.74	157 (100%)	-	-	2.0

* Using a mean refractive index increment found with many proteins.
** Assuming all the components have the same refractive index increment.

Experiments Using the Analytical Ultracentrifuge

The results are summarized in Table 3. All the virus samples, irrespective of type, contained a component with a sedimentation coefficient of 155-159 *S*. The experiments on the MEF_1 strain showed that the sedimentation coefficient was independent of the total concentration and the presence of the more slowly sedimenting components. A mean value of 156 *S* was obtained for all three polio types. In this calculation the data of Expts. 5 and 7 have been omitted, the former because it contained only a small amount of the fast sedimenting component, and the latter because the experiment was made in a separation cell.

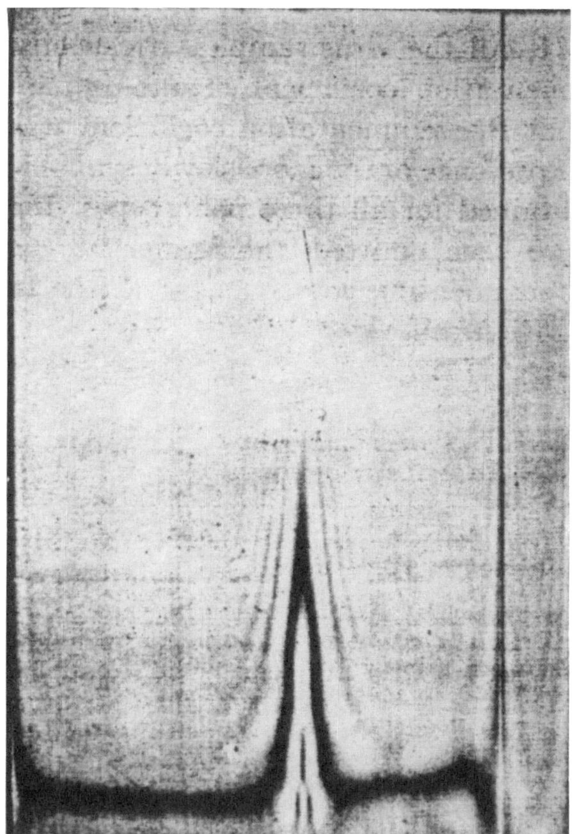

Figure 10. Sedimentation diagram of Type II virus, MEF_1 strain, Expt. 2 in Table 3. The exposure was taken after 35 min at 14,290 rpm Phaseplate angle 20°.

The sedimentation coefficients of the "slower" components were less consistent. The virus preparations appeared to contain in addition to the main component, two distinct groups, one of 70-95 *S*, Group II, and the other of 15-25 *S*, Group III. Owing to the short distance traversed by the slowest sedimenting component, no faith can be placed in the individual values in the 15-25 *S* group. The relative concentration of the 156 *S* component appeared to diminish as the total concentration of sedimented

Figure 11. Sedimentation diagram of Type I virus, Stockholm strain, Expt. 6 in Table 3. The exposure was taken after 52 min at 14,290 rpm Phaseplate angle 45°.

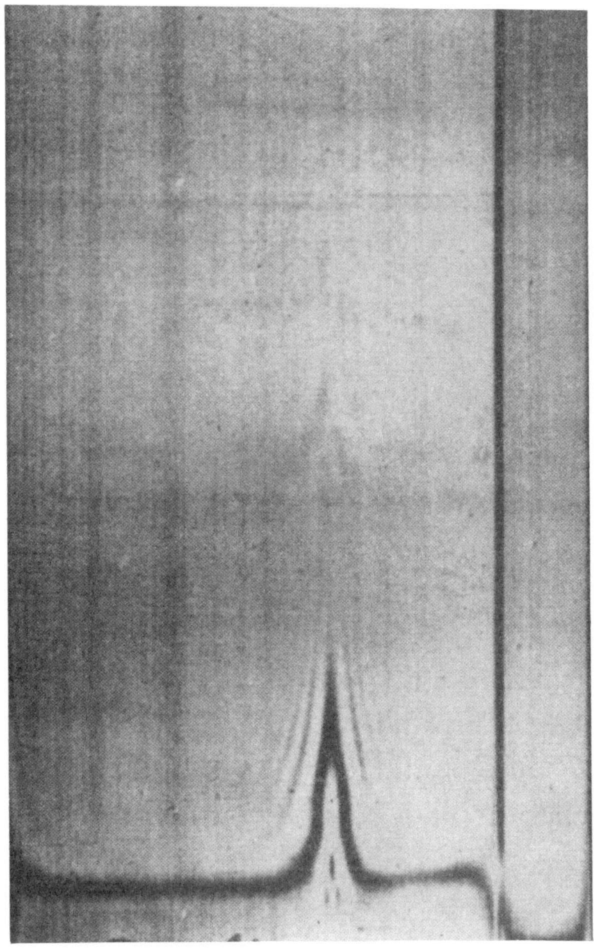

Figure 12. Sedimentation diagram of Type III virus, Saukett strain, Expt. 8 in Table 3. The exposure was taken after 26 min at 14,290 rpm Phaseplate angle 30°.

material was increased. The rest of the material was distributed equally between the two slower sedimenting components. Expt. 5 was an exception. This tendency might be due to a pronounced Johnston-Ogston effect[10], which would require that the sedimentation of the smaller particles is slowed down considerably in the presence of the faster sedimenting component. Alternatively, it is possible that an increasing proportion of the smaller components was formed during the preparation of the more concentrated virus suspensions. The latter assumption finds some support in experiments on lyophilized material, Ehrenberg and Polson[11]. When MEF_1 virus was freeze-dried, the 156 *S* component disappeared completely and 25-27% of the original material was found as a component with sedimentation coefficient 66-69 *S*, which may be considered as minimum for Group II.

As the sedimentation coefficient of component I appeared to be very constant, it was tempting to determine the diffusion coefficient of the component from the photographs. The spreading of the data is rather wide but the mean value for the MEF_1 and Stockholm I strains appeared to be 1.5 F (F=Fick unit = 1.10^{-7} cm^2/sec). This is comparable to the value 1.25 F that can be calculated from the sedimentation coefficient determined here and the particle mass 1.4 10^{-17}g and the partial specific volume of 0.64 reported by Schwerdt and Schaffer[12]. A single determination on a less concentrated sample of the Type III virus gave the value 2.0 F.

Distribution of Infectivity Amongst the Components in a Virus Preparation

In an early section of this work it was indicated that the infectivity of Type I virus was associated with components of two different sedimentation coefficients and in this respect it behaved like MEF_1 suckling mouse-adapted virus, Selzer and Polson[9]. There was, however, the important difference that in MEF_1 virus only approximately 1-10% of the infectivity was associated with the component of lower sedimentation coefficient, whereas with the Stockholm Type I virus the infectivity of the slower sedimenting component appeared to be considerably higher, as much as 10-30%. The Stockholm virus therefore, appeared to be admirably suitable for ultracentrifugal studies with the separation cell.

The preparation used in this study was identical with the material used in the analytical ultracentrifugation run 7 (in Table 3). The performance and results of this experiment can be seen in Fig. 13 and Table 4. In this experiment it has been assumed that nothing of the 156 *S* component was

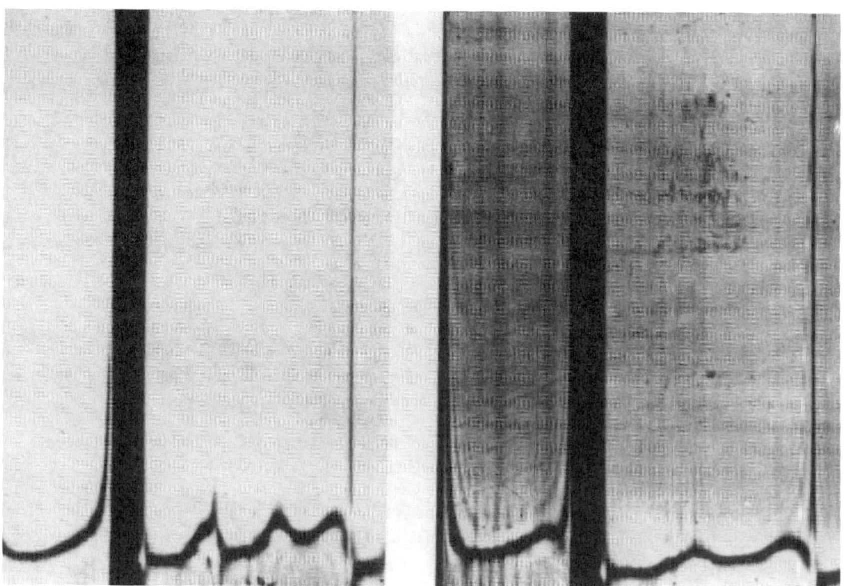

Figure 13. Sedimentation diagrams of Type I virus, Stockholm strain, in the separation cell. Exposure *a* taken after 39 min and *b* after 63 min at 14,290 rpm. Retardation was started 3 min after the latter exposure.

Table 4. Performed in the Separation Cell on the Virus Strain Stockholm Type I
The volumes of the top and bottom fractions were 0.39 and 0.26 ml respectively

Component	Concentration in mg/ml		
$S°_{20}$	Before Centrifugation	After centrifugation	
		in top fraction	in bottom fraction
159	1.35	-	3.37
81	0.80	0.32	1.52
16	0.85	0.73	1.02

left in the top compartment of the separation cell. A trace of this component might have been left behind, but even under unfavorable conditions, it could not have exceeded a concentration of 0.015 mg/ml, or 0.5% of the concentration of the same component in the bottom fraction. The infectivity test revealed $1.7 \cdot 10^8$ infective units per ml of the top fraction and 31.10^8 infective units per ml of the bottom fraction. If only the 156 *S* component were infective, the concentration of this in the top fraction must have been at least 2% of its concentration in the bottom fraction to account for all the infectivity (a maximum error of 0.25 in the log of the infectivity titre has been assumed). However, this could hardly be the case. It must therefore be assumed that the particles of both components 86 *S* and 156 *S* are infective. The relative infectivity of components 156 *S* and 86 *S* is thus estimated to be 1.3 on a weight basis. On a molecular or particle basis the 156 *S* fraction is approximately 4 times as infective as the 86 *S* portion.

This result confirms our previous observation with the help of the preparative Spinco that the infectivity is carried by particles of two different sedimentation coefficients. It can be mentioned in conclusion that with the gel-precipitin technique the inhomogeneous nature of the polio antigens can be shown very clearly, but this will form the subject of a separate paper[13].

DISCUSSION

The present work is confirmation of previous observations that the infectivity of poliomyelitis virus could be associated with components of different sedimentation coefficients. While previously the inhomogeneity of MEF_1 suckling mouse-adapted virus was shown by centrifugation in the preparative Spinco ultracentrifuge, this investigation centered around ultracentrifugation in the analytical Spinco. Strains of all three polio types were examined. Two strains of Type I virus were studied in the preparative centrifuge, these were 1423/53 Stockholm Type I and the Cincinnati virus of Sabin. From the ensuing sedimentation diagrams, it appeared that the Stockholm strain had a greater ratio of more slowly to more rapidly sedimenting virus than the Cincinnati virus. The Stockholm virus was therefore selected for further study in the analytical ultracentrifuge. It was found in these investigations that purified preparations of this strain contain three components, of sedimentation coefficients 156 *S*, 81-94 *S* and 16-20 *S*. The two faster sedimenting components have been shown by centrifugation in the separation cell to carry the infectivity, as was anticipated from the results obtained with the preparative centrifuge.

The sedimentation coefficients of the haemocyanins of *Helix pomatia* and *Caminella sincta* have been re-determined in the analytical Spinco centrifuge in view of their importance as reference substances in sedimentation studies of the smaller viruses. This was necessary on account of the discrepancies found between sedimentation coefficients as determined by the oil turbine and the analytical Spinco ultracentrifuges. It was found that the sedimentation coefficient of *H. pomatia* was roughly 10% lower than the value quoted by Svedberg and Pedersen[6]. When this lower figure was used, sedimentation coefficients were calculated from the data of runs in the preparative centrifuge, and values were obtained which differed very little from those of the analytical runs.

Purified preparations of the MEF_1 and Saukett strains have also been investigated by analytical ultracentrifugation. The former showed some inhomogeneity, but in the latter no sign of inhomogeneity was evident. The sedimentation coefficient of the main components of the three types of polio virus did not differ materially, and values ranging between 150 and 159 *S*, with an average value of 156 *S*, were obtained. This value is almost identical with the figure 154 *S* obtained by Schwerdt and Schaffer[2] on a purified preparation of MEF_1 virus.

The phenomenon of the sharing of infectivity by particles of two different sizes might be related to the observation of Taylor and McMormick[14] that 0.25% glycerol can cause a disruption of purified polio virus particles whereby particles of smaller diameter are produced. From their measurements on electron micrographs it would appear that these particles were approximately 20 mμ in diameter. A definite proof of the existence of these smaller particles in the dilute glycerol solution is, however, not yet available. Such a particle dimension would correspond roughly to a sedimentation coefficient of about 90 *S*, if the partial specific volumes lies between 0.60 and 0.70 ml/g.

SUMMARY

Preparation of types I and II of polio virus have been found by sedimentation analysis to contain three distinct components with sedimentation coefficients of 156 *S*, 70-95 *S* and 15-25 *S* respectively. The Stockholm strain of Type I virus has a greater ratio of more slowly to more rapidly sedimenting material than the other polio virus types investigated. A diffusion coefficient of 1.5 F has been estimated for the 156 *S* component. Infectivity tests following separation centrifugation of one sample indicate that the two faster components are both infective. The Type III virus appeared homogeneous in the ultracentrifuge.

ACKNOWLEDGMENTS

The authors would like to acknowledge their gratitude to Drs. Wrange and Melén for placing large volumes of polio tissue-culture fluid at their disposal and for technical help during the execution of this work. Able technical assistance was given by Mr. N. Ekman in preforming the ultracentrifuge experiments. To Dr. Sjöstrand grateful acknowledgments are made for permission to use the electron micrographs of the purified virus preparations in this publication. One of us (A.P.) is indeed very grateful to Professor Sven Gard of the Karolinska Institute for financial aid during his stay in Stockholm (August 1956 to April 1957) without which this work would not have been possible. Thanks are further due to the Knut and Alice Wallenberg Foundation for financial support.

REFERENCES

1. G. Selzer and A. Polson, Biochim. Biophys. Acta, *15*, 251 (1954).
2. A. Polson and A. Linder, Biochim. Biophys. Acta, *11*, 199 (1953).
3. A. Polson and J.W.F. Hampton, J. Hyg., *55*, 344 (1957).
4. A. Polson and T.I. Madsen, Biochim. Biophys. Acta, *14*, 366 (1954).
5. G.L. Miller and R.H. Golder, Arch. Biochem. Biophys., *36*, 249 (1952).
6. T. Svedberg and K.O. Pedersen, The Ultracentrifuge, The Clarendon Press, Oxford, 1940.
7. S. Brohult, Thesis, University of Uppsala, 1940.
8. A. Ehrenberg, Acta Chem. Scand. *11*, 1257 (1957).
9. A. Polson and G. Selzer, Proc. Soc. Exptl. Biol. Med., *81*, 218 (1952).
10. J.P. Johnston and A.G. Ogston, Trans. Faraday Soc., *42*, 789 (1946).
11. A. Ehrenberg and A. Polson, Biochim. Biophys. Acta, *28*, 442 (1958).
12. C. Schwerdt and F.L. Schaffer, Ann. N.Y. Acad. Sci., *61*, 740 (1955).
13. A. Polson, A. Ehrenberg and R. Cramer, Biochim. Biophys. Acta, *29*, 622 (1958).
14. A.R. Taylor and M.J. McCormick, Yale J. Biol. and Med., *28*, 589 (1956).

16. Specific Precipitin Reactions of the Virus of Poliomyelitis in Gels*

ABSTRACT

By means of precipitin tests, preparations of Type I polio virus have been found to contain at least three antigenically active components. The three precipitin bands have tentatively been correlated to the three distinct components observed in the ultracentrifuge. Type II also contains more than one antigen but the secondary bands are weaker. A lyophilized sample produced a precipitin pattern very similar to that formed by the untreated virus, in spite of the drastic changes in virus infectivity and particle size following this treatment. The Type III virus appeared to be antigenically homogeneous. No cross reactions could be observed between the different virus types and immune sera.

INTRODUCTION

Precipitin reactions in gels according to the technique of Oudin[1], Ouchterlony[2] and Oakley and Fulthorpe[3] are becoming increasingly important, not merely as a means of identifying antigens, but also as an analytical tool for determining the minimum number of antigenic components in mixtures of antigens. Furthermore, in view of the minute amount of detectable antigen and the simplicity of the technique involved, it is in addition becoming a valuable supplement to electrophoresis and ultracentrifugation.

*From: A. Polson, A. Ehrenberg and R. Cramer, Biochim. Biophys. Acta, *29*, 622-629 (1958).

Most of the published work employing this technique has made use of antigens of bacterial origin, and it is only relatively recently that virologists have applied this procedure to their problems. The influenza group of viruses has been studied by Jensen and Francis[4] by this technique. Gispen[5] applied it to the pox group, Bodon[6] examined foot-and-mouth-disease virus and Brown and Crick[7] used the method for differentiating between vesicular stomatitis and foot-and-mouth-disease virus.

However useful the technique may appear great caution is demanded in the interpretation of the results from such gel precipitin tests. Secondary effects frequently appear, especially when working with too high concentrations of the reactants and in observations over long periods of time. Jennings[8] has shown that double precipitin bands may form from a single antigen under certain conditions, and Boerma[9] has indicated that it is only when the antigen and antibody are present in equivalent amounts that a sharp precipitin band may be expected. When the antigen is in excess the precipitin bands tend to grow out in the direction of the antibody, and when the antibody is in excess the converse is true. In the immediate vicinity of the equivalent zone there is a tendency for the precipitin bands to form doublets. One of the authors (A.P.) has noticed a similar effect in precipitin reactions in gels with the hemocyanin of *Jasus lalandi*. These doublets may easily be mistaken for antigenic inhomogeneity. No adequate explanation for this phenomenon is available at the moment.

In the present work, the gel precipitin method of Oudin as modified by Oakley and Fulthorpe[3] was applied to representative strains of the three types of poliomyelitis virus, which had been purified from large volumes of tissue culture fluid by the method of Polson and Hampton[10].

MATERIALS

Virus Strains

The following strains of poliomyelitis virus were investigated: 1423/53 strain of Type 1 (Stockholm strain); MEF_1 strain of Type II; Saukett strain of Type III.

Immune Sera

Both guinea-pig and monkey immune sera prepared against the three virus types were used in this study.

Table 1. Gel Precipitin Tests on Purified Poliomyelitis Virus Suspensions

Experiment	Material	Neg. log. $TCID_{50}$	Titre in gel
a	Type I	10.6	1/200 homologous antiserum, monkey
b	Type II	9.0	1/1000 homologous antiserum, monkey
c	Type III	9.0	1/1000 homologous antiserum, monkey
d	Lyophilized II	2.5	1/1000 homologous antiserum, monkey
e	Top fraction I u.c.*	7.25	1/1000 homologous antiserum, monkey
f	Bottom fraction I u.c.*	8.5	1/1000 homologous antiserum, monkey
g	Type I	9.6	1/20 Guinea-pig homologous antiserum
h	Type I	9.6	1/20 Type II antiserum guinea-pig
i	Type I	9.6	1/20 Type III antiserum guinea-pig
j	Type I heated		1/20 Type I antiserum guinea-pig
k	Type I heated		1/20 Type II antiserum guinea-pig
l	Type I heated		1/20 Type III antiserum guinea-pig
m	Type II	9.5	Normal monkey serum
n	Type II	9.5	1/1000 Type III antiserum, monkey

* u.c. = ultracentrifuge
See Text

Recording of the Precipitin Bands

Several attempts at recording the precipitin band photographically failed. These efforts were abandoned in favour of photometric recording with the aid of a Hilger-Watts microphotometer. This method seems to be very powerful. However, with the instrument used the resolving power was somewhat limited because of the depth and geometry of the object (test tubes with about 10 mm diameter) and incomplete parallelism of the light beam. This might easily be improved if an instrument is constructed especially for the purpose of measuring precipitin bands.

The investigations in this study are summarized in Table 1.

RESULTS

In the following figures photometric diagrams of the gel precipitin tests are presented. The results obtained with the separate strains of virus are given together. The antigen menisci are indicated on the left and the antibody menisci on the right hand side of the diagrams.

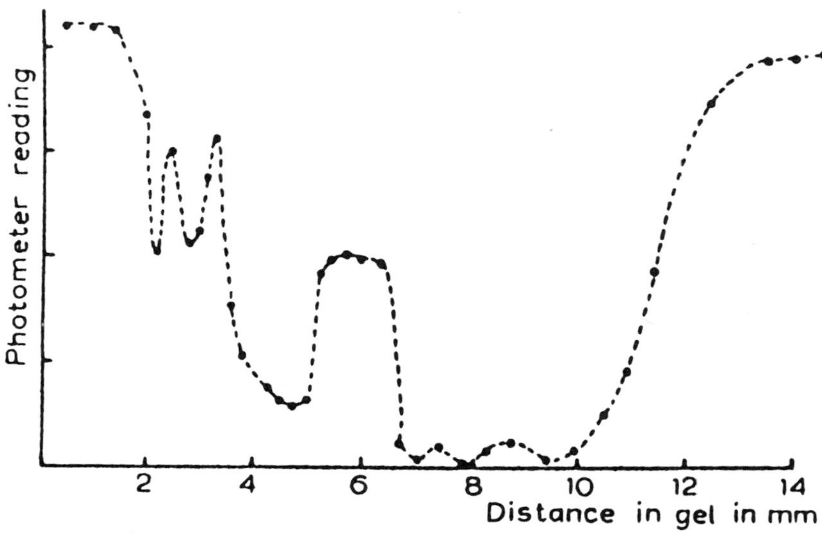

Figure 1. Photometric diagram of gel precipitin band of purified Type I polio virus with its homologous monkey immune serum. Expt. a.

Type I Virus

Experiment a.

In Fig. 1 appears a diagram obtained when purified Type I virus of high titre was diffused against its homologous immune serum obtained from monkeys. There were at least three precipitin bands; two, the first and second, appeared after 36 h incubation at 37° and the third became visible after 6 days. The broad band developed into a doublet after an incubation period of approximately 30 days. The bands in the doublet were too close together to be resolved by the microphotometer but they could be identified by unaided eye.

Experiments c and f.

In these experiments it was attempted to correlate the bands in the gel-precipitin test with the three components defined by ultracentrifugation of the virus suspension in the analytical Spinco ultracentrifuge, (Polson, Ehrenberg and Cramer[11]). For this purpose a purified virus suspension was centrifuged in the separation cell. Care was taken to sediment the component of sedimentation coefficient 156 S out of the top part of the separation cell and to leave a portion of the 81-94 S and the bulk of the 16-20 S components in the top section of the separation cell.

In Figs. 2a and 2b are recorded the photometric diagrams of the precipitin bands of the material recovered from the top and bottom compartments respectively. These precipitin bands were recorded after 12 days' incubation and the diagrams presented are by no means the final result of the test. During the ensuing 4-day period, bands α and β became visibly stronger. The rest of the bands did not similarly increase in intensity as judged by the unaided eye. The preparation used in these experiments was different from the one used in experiment a.

Experiments g, h, i, j, k and l.

In these experiments attempts were made to determine whether there were any cross reactions of Type I virus with the immune sera against Types II and III and whether heating of the virus of Type I at 56° for 30 min caused the liberation of any antigen which might react with the antisera of the other types of poliomyelitis virus. In this series of tests immune guinea-pig sera were used.

In Fig. 3 the diagram obtained with Type I virus against homologous guinea-pig immune serum is given. The first two bands made their appearance after an incubation period of approximately 6 days and the third showed up clearly after 14 days. In Fig. 4 is given the diagram of the identical virus heated to 56° for 30 min and tested in exactly the same manner against the same immune serum. The bands were formed after approximately the same incubation period as for the unheated antigen, but he diagram obtained was different in that the intermediate band was markedly broader. No definite precipitin bands were obtained when the virus in both the unheated and heated states was diffused against the heterologous sera. A typical diagram is shown in Fig. 5.

Type II Virus

Experiment b.

In Fig. 6 a diagram is given of purified Type II virus tested against its homologous monkey immune serum. This is the final diagram obtained

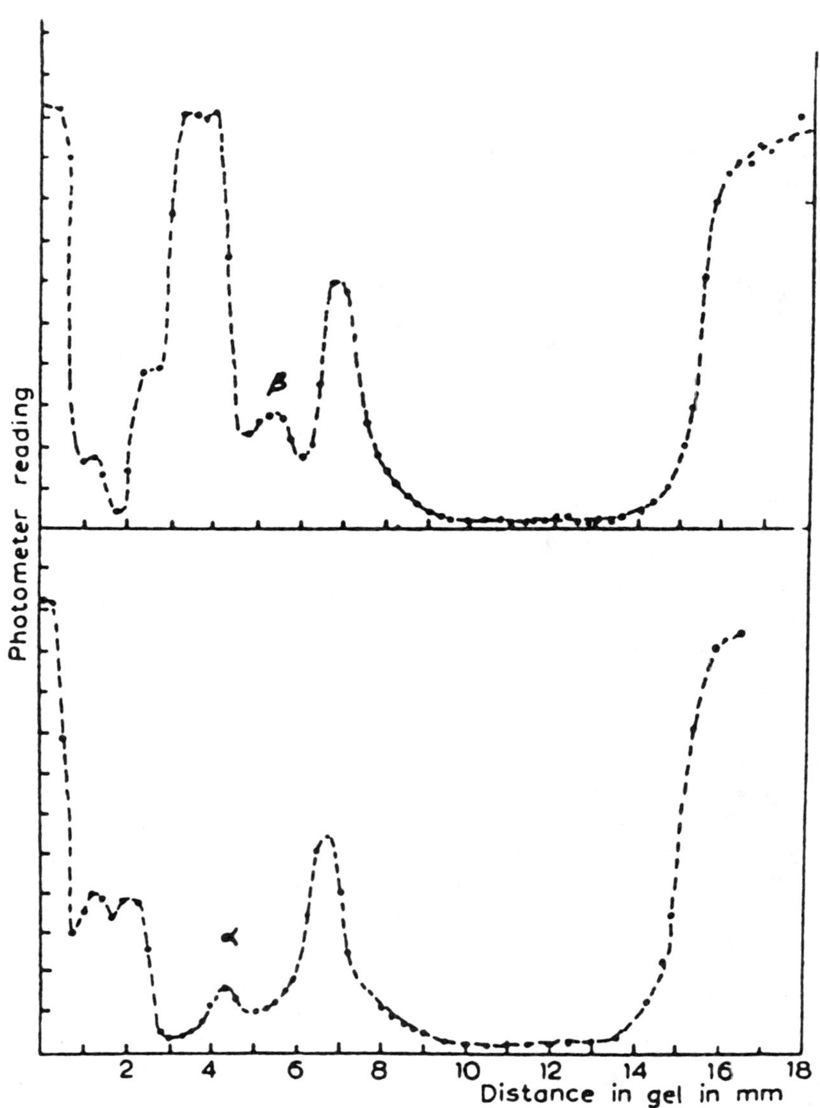

Figure 2a and b. Photometric diagrams of the precipitin bands of the material recovered from the top and bottom compartments after centrifugation of purified Type I virus in the separation cell of the Spinco analytical ultracentrifuge. Expt. e and f.

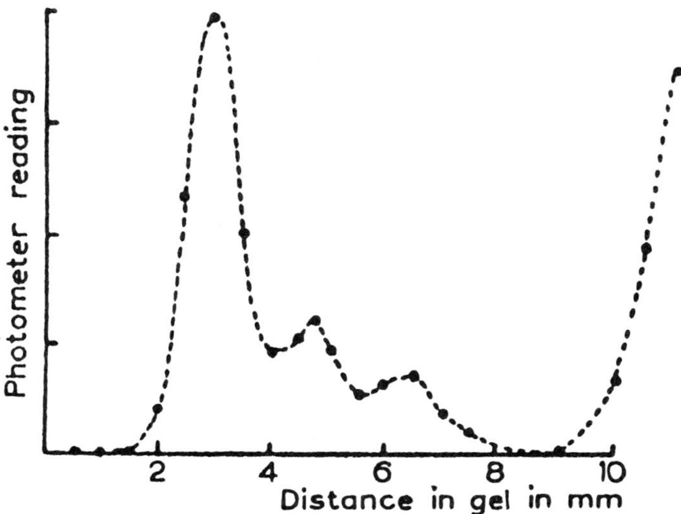

Figure 3. Photometric diagram of the precipitin bands of purified Type I polio virus against homologous guinea pig immune serum. Expt. g.

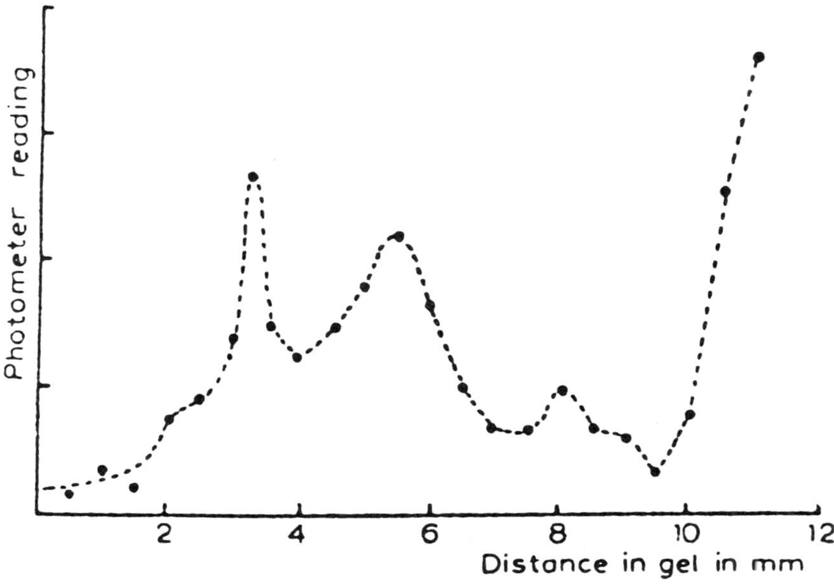

Figure 4. Photometric diagram of the precipitin bands of purified Type I virus heated at 56° for 30 min and tested against its homologous guinea-pig immune serum. Expt. j.

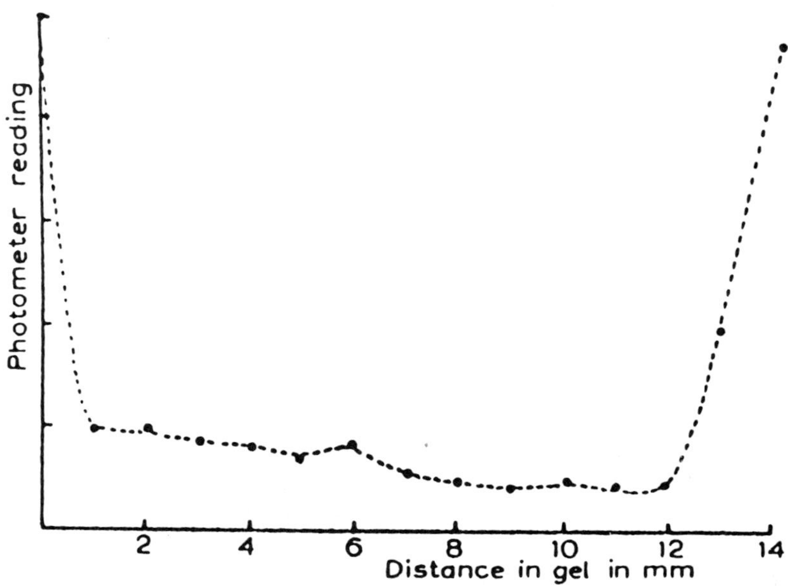

Figure 5. Photometric diagram of the results of a gel precipitin reaction of unheated Type I virus against Type III guinea-pig immune serum. Expt. i.

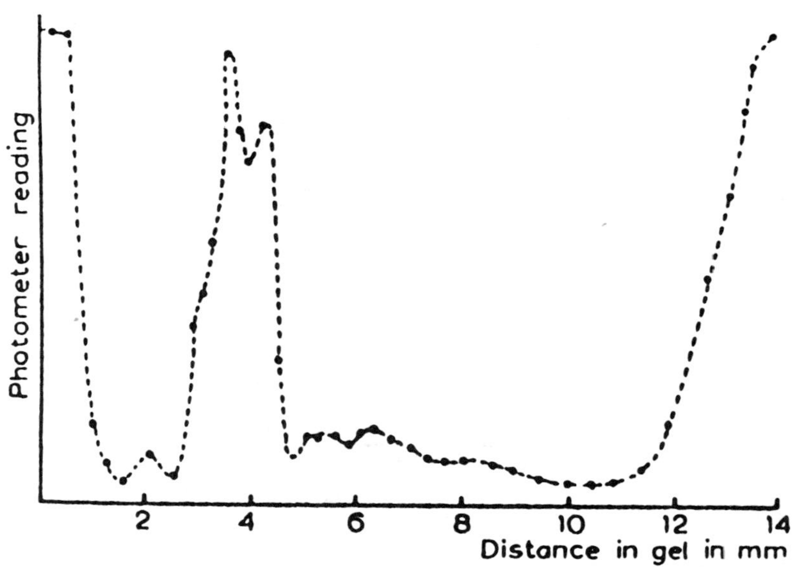

Figure 6. Photometric diagram of the result of a gel-precipitin reaction of purified Type II polio virus against its homologous monkey immune serum. Expt. b.

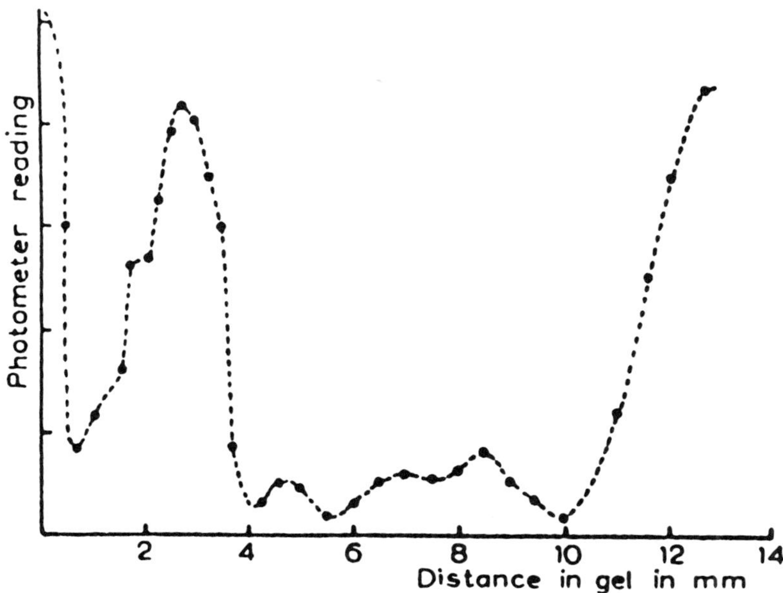

Figure 7. Photometric diagram of the result of a gel-precipitin reaction of lyophilized Type II polio virus against homologous monkey immune serum. Expt. d.

after 23 days' incubation at 37°. After 2 days' incubation the main band appeared and a second band started to develop just below it on the 13th day. This was followed by a weak band immediately above it on the 20th day. At the same time two broader bands formed lower down in the tube.

Experiment d.

A lyophilized preparation of Type II virus which was also analyzed in the Spinco ultracentrifuge and examined in the electron microscope[14] was used in this precipitin test. After 3 days' incubation a broad band appeared in the gel which became progressively heavier until the 14th day when the broad band appeared to reform into two heavy bands separated by a very short distance from each other. In addition, faint broad bands developed lower down in the gel during this interval. In Fig. 7 the photometric recorded of the precipitin bands is given.

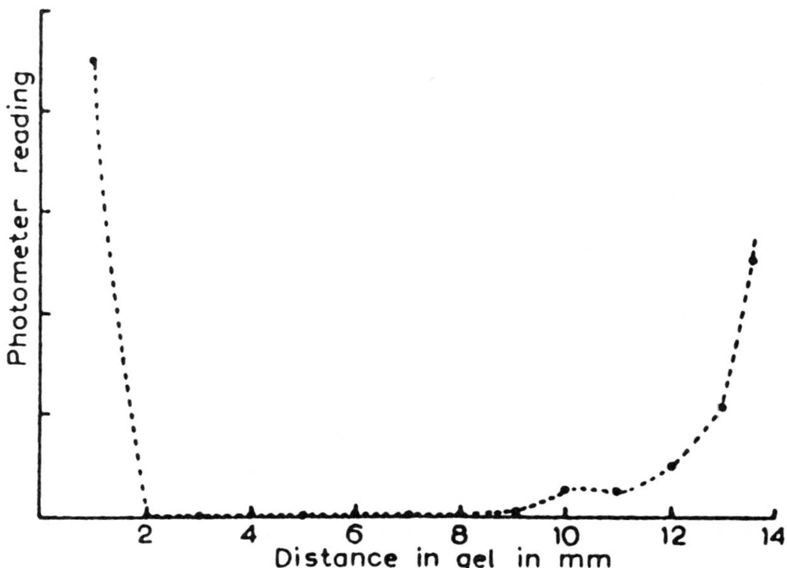

Figure 8. Photometric diagram of a control test where Type II virus was tested against normal monkey serum. Expt. m.

Experiment m.

In Fig. 8, the result of a control experiment is given. In this test Type II virus was diffused against normal monkey serum. The complete absence of precipitin bands is obvious from an examination of the diagram.

Experiment n.

In this it was attempted to show up cross precipitin reactions of Type II virus with Type III antiserum. However, no evidence of any cross reaction was obtained.

Type III Virus

Experiment c.

In Fig. 9, an early diagram obtained when purified Type III virus was tested against its homologous monkey serum is given. The main precipitin

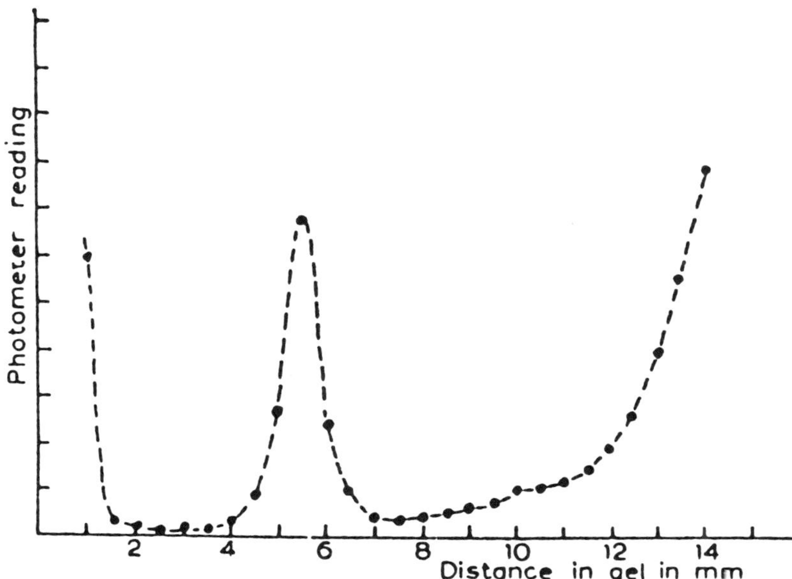

Figure 9. An early photometric diagram of the precipitin band of purified Type III polio virus tested against its homologous monkey immune serum. Expt. c.

band made its appearance after 36 h incubation and became progressively stronger until the 21st day. Photometric recordings made of the test during this period failed to show up any additional precipitin. However, after 23 days of incubation two new bands appeared, one immediately above and the other immediately below the main precipitin band. Two faint broad bands were formed lower in the tube during the subsequent period. The precipitin bands are shown in Fig. 10.

In a control experiment of Type III virus against normal monkey serum, no precipitin bands developed during the observation period of 57 days.

INTERPRETATION AND DISCUSSION

Experiments with Type I Virus

The precipitin test with this virus revealed the presence of at least three antigenic compounds in the purified preparation. The main band appeared

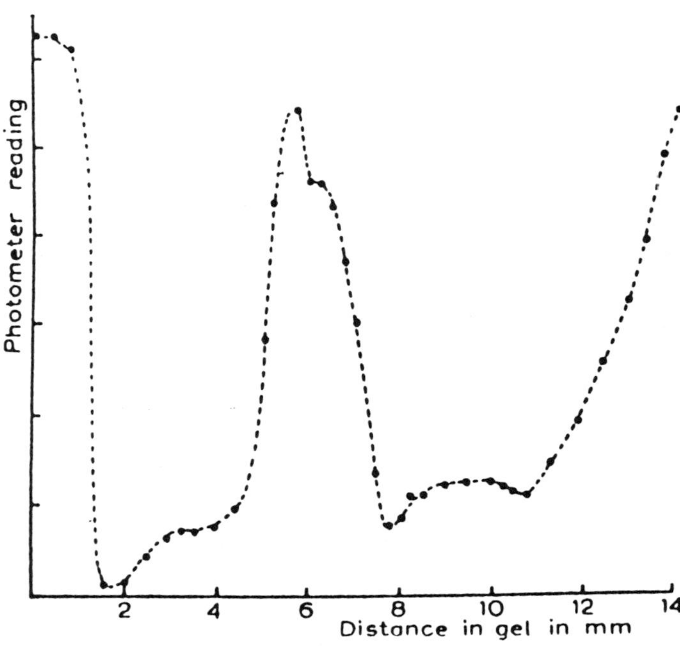

Figure 10. Final photometric diagram of the precipitin bands of purified Type III polio virus tested against its homologous monkey immune serum. Expt. c.

as a "close doublet". This, however, should not be regarded as indicating two antigenic components in the main band, but rather as showing that he band-doubling and broadening effects as noticed by Boerma[9] and Jennings[8] are active. If this interpretation is correct the three antigens observed here probably correspond to the three components shown to be present in ultracentrifugation analysis of this strain[11]. An attempt was made to correlate the precipitin bands with the components observed in the ultracentrifuge. To do this, gel-precipitin tests were made on the material recovered from the top and bottom sections of the separation cell after the main component 156 S had been centrifuged past the separation section of the cell, and until a portion of the 81-94 S component passed into the bottom compartment. This manoeuvre indicated that he main precipitin band (doublet) was formed by the 156 S component. The 16-20 S component is probably responsible for the peak closet to the serum meniscus since it shows very little difference in intensity and in position relative to the antigen meniscus in the tests made on the contents of the top and bottom sections of the separation cell. This result can be expected from the relatively low sedimentation constant of this component. The third component of the virus preparation (the 81-94 S component) is probably the one responsible for the band which occupies the intermediate position between the bands of the 156 S and 16-20 S components. This deduction is based on the following argument.

According to the theory of the gel precipitin test the position relative to the antigen meniscus where a band is formed is dependent on the concentrations of the antigen and/or antibody. With constant antibody concentration the distance from the antigen meniscus to the position where the precipitin band is formed is proportional to the logarithm of the antigen concentration. If we now assume that bands α and β were formed by the same antigen, the change in position relative to the antigen meniscus might well be expected from the decrease of its concentration as a result of centrifugation in the separation cell. This effect has been verified by one of us (A.P.) in experiments on precipitin reactions with the Brunhilde strain of Type I poliomyelitis virus and on haemocyanin. By plotting the distance of the precipitin bands from the antigen menisci against the logarithm of the virus or haemocyanin concentration a straight line is obtained.

Furthermore, several authors have shown that there is a relationship between distance of migration or occurrence of a precipitin band and the size of the antigen particle. Thus Franlin and Kunkel[12] have shown that the 7 S component of γ-globulin always forms a precipitin band with its antibody further away from the antigen source, while the band of the 19 S component follows well behind this. Also Neef and Becker[13] have calculated that the rate of migration of a precipitin band in the Oudin test is a function of the square root of the diffusion constant. This was shown experimentally in precipitin test with the following proteins: egg albumin, human serum albumin, bovine γ-globulin, *homarus americanus* haemocyanin and thyroglobulin.

In view of this evidence, the three bands found for Type I polio are probably formed by the 156 S, the 81-94 S and the 16-20 S components in order of the distance of the precipitin band from the antigen source.

No precipitin bands were formed between Type I virus and the immune sera of guinea pigs immunized against Type II or Type III virus, nor did the Type I virus, heated to 56° for 30 min, show any reactions with these sera.

Heated Type I virus produced a diagram with its homologous serum which was appreciably different from that produced by the unheated material. As yet no explanation is offered for this observation.

Experiments with Type II Virus

The main band which formed in gel precipitin tests with Type II virus showed a close doublet finally, similar to that found with Type I virus. Weaker bands developed lower down in the tube, indicating that this virus preparation also contained other antigenically active substances. A lyophilized preparation of the virus produced a precipitin pattern very similar to that formed by the untreated virus. This happened in spite of the drastic reduction in virus infectivity from a titre of neg. log. 8.5 before freeze-drying to 2.5 after this treatment and in spite of the dissociation of the virus as observed in the ultracentrifuge and in the electron microscope, (Ehrenberg and Polson[14]). The present preparation showed two components in the ultracentrifuged sample. These had sedimentation coefficients of 69 and 15 S respectively.

It was shown that Type II virus did not produce any precipitin bands with normal monkey and Type III monkey immune sera.

Experiments with Type III Virus

This virus preparation appeared to be antigenically homogeneous during an observation period of 23 days. However, as was the case with the other types of polio virus, doublet formation occurred after this time-interval. This virus preparation was homogeneous by ultracentrifugation analyses when the precipitin test was initiated, (Polson, Ehrenberg and Cramer[11]).

In conclusion, it would appear from the results of the precipitin reactions that the preparations of Type I virus are relatively more inhomogeneous than the preparations of the other polio types. Analogous results have been obtained by sedimentation analysis[11]. As the same technique of purification has been used with this virus as with the Types II and III agents it is concluded that the particles of the Type I virus used in this work might have a greater tendency to dissociate into smaller components of 81-94 S and 16-20 S than have the other polio types.

Similar work has recently been published by Le Bouvier, Schwerdt and Schaffer[15] and Le Bouvier[16] in which the three types of poliomyelitis virus were studied by the gel-precipitin technique. These authors also found that no precipitates formed by heterotypic virus-serum combinations. Two bands of precipitate developed when purified preparations of all three types were tested against their homologous antisera.

ACKNOWLEDGEMENTS

One of us (A.P.) wishes to express his gratitude to Professor S. Gard, of the Karolinska Institute, Stockholm, for financial aid during his stay in Stockholm (August 1956 to April 1957). The authors are also grateful to the staff of the State Bacteriological Laboratory, Stockholm, for the generous supply of virus used in this work.

REFERENCES

1. J. Oudin, Ann. Inst. Pasteur, *lxxv* 30, 109 (1948).
2. O. Ouchterlony, Ark. Kemi Mineral. Geol., *26* B, 1 (1948).
3. C. L. Oakley and A. J. Fulthorpe, J. Pathol. Bacteriol., *65*, 49 (1953).
4. K. E. Jensen and T. Francis, J. Immunol., *70*, 321 (1953).
5. R. Gispen, J. Immunol., *74*, 135 (1955).
6. L. Bodon, Acta Vet. Acad. Sci. Hung., *5*, 157 (1955).
7. F. Brown and J. Crick, Nature, *179*, 316 (1957).
8. R. K. Jennings, J. Bacteriol., *67*, 565 (1954).
9. F. W. Boerma, J. Pathol. Bacteriol., *lxxii*, 515 (1956).
10. A. Polson and J. W. F. Hampton, J. Hyg., *55*, 344 (1957).
11. A. Polson, A. Ehrenberg and R. Cramer, Biochim, Biophys, Acta, *29*, 612 (1958).
12. E. C. Franklin and H. G. Kunkel, J. Immunol., *78*, 11 (1957).
13. J. C. Neef and E. L. Becker. J. Immunol., *78*, 5 (1957).
14. A. Ehrenberg and A. Polson, Biochim. Biophys. Acta, *28*, 442 (1958).
15. G. L. le Bouvier, C. E. Schwerdt and F. L. Schaffer, Virology, *4*, 590 (1957).
16. G. L. le Bouvier, J. Exptl. Med., *106*, 661 (1957).

17. Purification and Aggregation of Influenza Virus by Precipitation with Polyethylene Glycol*

ABSTRACT

Influenza virus may be purified and rendered free of extraneous proteins by precipitation and aggregation with polyethylene glycol at polymer concentrations of 1 to 4%. The precipitated virus is superior antigenically to the virus in monomeric and in the ether dissociated forms.

When the virus is precipitated at polyethylene glycol concentrations of 5% and higher the virus is not aggregated and is associated with extraneous protein which co-precipitates with the infectious agent.

INTRODUCTION

In an earlier communication, Polson *et al*[1] reported that influenza virus could be precipitated by polyethylene glycol m.w. 6,000 (p.e.g.) at polymer concentrations from 1% and higher. The virus precipitates removed from suspension by centrifugation at low speed (2,000 rpm for 30 min) were only sparingly soluble in electrolyte solutions as judged by haemagglutination (HA) titres and could consequently be freed of most extraneous matter by washing in appropriate buffer solutions. It was further shown that the virus aggregates were antigenically more effective in stimulating an antibody

*From: A. Polson, Prep. Biochem., *4*: 435-456 (1974).

response in guinea pigs as compared to the monomeric unaggregated form of the virus[1].

The ability of p.e.g. to aggregate the virus at low (4%) polymer concentration is in apparent contrast to the observation of Kanarek and Tribe[2]. Their use of p.e.g. concentration of 7% or higher indicated that the virus could be recovered in the monomeric form by redissolving the precipitate in the appropriate buffer. The following contribution offers a satisfactory answer to the apparent conflicting observations of Kanarek and Tribe[2] and Polson *et al.*[1]

Virus

The virus used in this study was obtained from allantoic fluid harvested from eggs infected with the X31 recombinant strain of influenza.

Polyethylene Glycol

Polyethylene glycol (p.e.g.) m.w. 6,000 as supplied by Shell Chemicals, Cape Town, was used as precipitant. In order to increase the rate of dissolution of the polymer it was finely ground in a "Waring" blender and added as a powder to the infected fluid.

Two methods of precipitation were applied to the virus. They were termed "incremental" and "aliquot" precipitation respectively.

"Incremental" Precipitation

The precipitation of the virus by p.e.g. is demonstrated by a step-wise gradient method (incremental precipitation). In this technique the polymer was added in varying increments and the resultant precipitate centrifuged off and redispersed in a volume of phosphate buffered saline (PBS), pH 7.0, equal to that of the original allantoic fluid. Thus p.e.g. was added and dissolved in the infected allantoic fluid to an initial concentration of 0.6 g/100 ml. After the mixture was left on the bench at room temperate for 30 min for the approximate completion of the precipitate, the solution was centrifuged at 1000g for 30 min. The pellets were well drained of liquid prior to redispersion. The concentration of p.e.g. in the clear supernatant (SNF) was then raised to 1% by the addition of a further 0.4 g/100 ml. The precipitate which formed was again centrifuged off and dispersed in a volume of PBS equivalent to the original. Thus the precipitates formed at 0.6, 1, 2, 3, 5 and 7% p.e.g. were dispersed and stored together with their corresponding SNF's.

"Aliquot" Precipitation

Pulverized p.e.g. was dissolved in equal portions of infected allantoic fluid so that the final polymer concentrations were 0.6, 1, 2, 4, 5, 6 and 7.5%. The reactions were allowed to proceed for 3 hr at 4°C after which the precipitates were centrifuged off at 1000g for 30 min. After drainage of most of the p.e.g. solution from the centrifuge tubes, the precipitates were dissolved in volumes equivalent to that of the original with 0.066 M phosphate buffer at pH 7.0. Haemagglutination activities were determined on the respective SNF's and on the redispersed precipitates.

Concentration and Purification by Thin Layer Centrifugation

In this study, purified influenza virus in its monomeric state was often needed. To obtain the virus in this state, it was centrifuged from suspensions at relatively low rotor velocity into the receiving cavity of the thin layer centrifuge rotor (TLR) (Polson and Stannard[3]). By this technique of centrifugation the infective particles were recovered as a concentrated suspension of monomeric particles and not as a tightly packed pellet as occurs with conventional ultracentrifugation. Batches of 100 ml infected allantoic fluid were centrifuged at 6,000 rpm (5000g) for periods of 90 min. The pool of the concentrates was dispersed in 100 ml PBS and centrifuged off for a final "wash" to free the preparation of proteins of the allantoic fluid. The final virus concentrate had a volume of approximately 2-3 ml.

Analytical Ultracentrifugation

Analytical ultracentrifugation runs were made in the Model E. The rotor was accelerated to 8,766 rpm and the experiments were done at 20°C. Photographs of the sedimenting boundaries in which the schlieren system was used were taken at 8 min intervals.

Protein Assay

The protein contents of (a), the allantoic fluids after removal of the virus by various means and (b) virus suspensions obtained from infected allantoic fluid by different means, were determined by the standard biuret method.

Haemagglutination (HA)

Haemagglutination assays were made on the virus preparations whenever necessary. The titration endpoints were read where 100% agglutination of the red cells occurred.

Electron Microscopy

Electron micrographs were made of the virus purified by thin layer centrifugation and by precipitation with 1% p.e.g. The preparations were negatively stained with phosphotungstic acid and examined in an Elmiskop 1A.

IMMUNOGENICITY OF THE CONCENTRATES

In the previous communication[1], evidence was obtained that following intramuscular injection of guinea pigs the virus aggregated by treatment with 4% p.e.g. was more immunogenic than equal amounts of virus in the monomeric form or in the dissociated state. However, it was noted that the observations were made on too few animals. The work was repeated and a larger number of guinea pigs per test sample was used. The immunizing dose per animal used in this experiment was considerably larger than that previously used.

Preparation of the Vaccine

Infected allantoic fluid was concentrated tenfold by perosmosis against p.e.g. m.w. 20,000. The concentrate was divided into three equal portions. One portion was diluted in PBS and subjected to centrifugation in the TLR at 5000g for 90 min. The virus concentrate obtained in this manner was contained in 2 ml and is referred to as the "monomeric" form. The second portion of the concentrated infected allantoic fluid was treated with 4% pulverized p.e.g. and the precipitated virus centrifuged off at 1000g for 20 min. The virus precipitate was dispersed in 2 ml PBS and was referred to as the "aggregated" form. The virus in the third portion was precipitated with 4% p.e.g., centrifuged off, dispersed in 2 ml PBS and shaken briskly in 4 ml ether for 90 min in the cold. This preparation was referred to as "ether" dissociated virus.

To the three concentrates, formaldehyde was added to a final concentration of 0.1 g/100 ml and the mixtures held at room temperature for

24 hours. The three preparations thus obtained were termed:

(a) monomeric influenza;
(b) aggregated influenza;
(c) ether dissociated influenza

Immunization of Guinea Pigs

The three vaccines were injected into 3 groups of 10 guinea pigs respectively, each animal receiving 0.2 ml intramuscularly. The animals were bled by heart puncture 13 days after inoculation. The sera were subjected to heat treatment at 56°C for 30 min and portions of each were set aside for haemagglutination inhibition titration. These aliquots were termed "untreated". The remaining portions of the sera were treated with receptor site destroying enzyme (RDE) for 24 hours. These were termed "treated".

RESULTS

Precipitation with p.e.g.

The results obtained in experiments in which the virus was precipitated with p.e.g. by the incremental method from infected allantoic fluid are presented in Fig. 1. The solid line represents the reduction of HA activity in the supernatant fluid and the dotted line the HA activities of the redispersed precipitates in amounts of PBS equal to that of the original untreated allantoic fluid. The broken line represents the differential reduction of the HA activities at the different levels of polymer concentration. It appears from the data that the sum total of the recovered HA activities following redispersion of the precipitates was 25 to 30% of that originally present in the infected allantoic fluid, despite the removal of more than 90% of the activity at 4% polymer concentration.

In Fig. 2 the redispersed precipitates obtained with p.e.g. at concentrations 1 to 4% (aliquot experiment) showed HA activities of approximately 5% of the original despite the almost complete removal of the HA activity from the SNF's over the range. Opposed to this, the redispersed precipitates at 5, 6 and 7.5% polymer indicated full recovery of HA activity.

Concentration by TL Centrifugation

When influenza virus is concentrated by thin layer centrifugation the recovery of virus is usually quantitative after the initial concentration. The

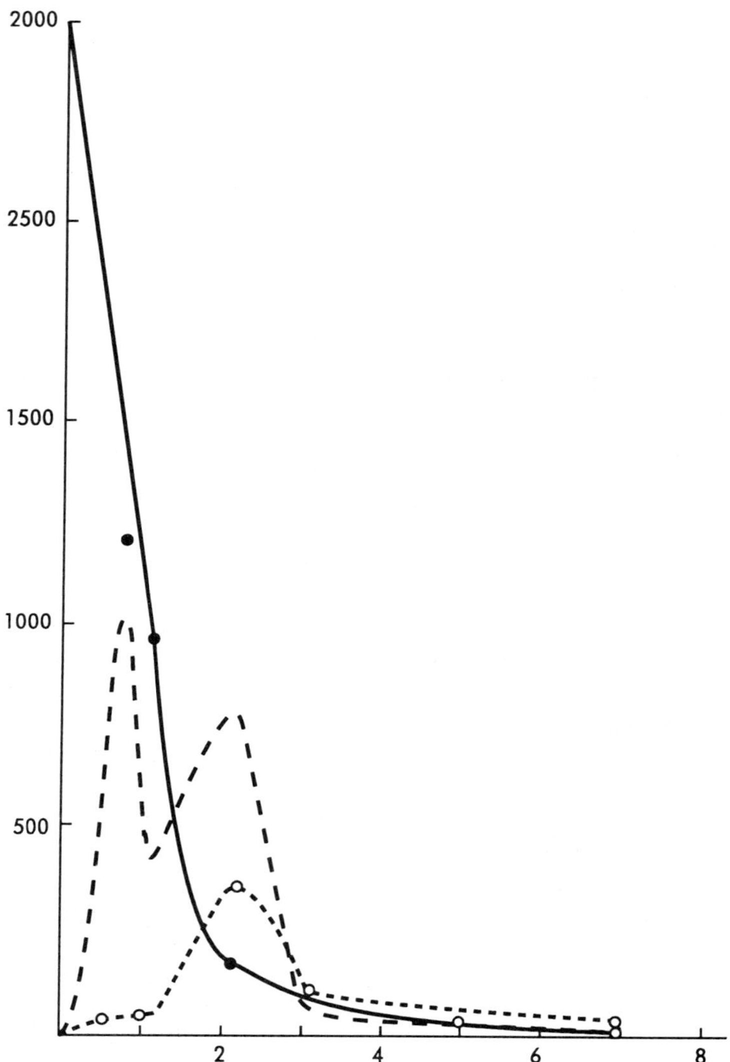

Figure 1. "Incremental" precipitation of X31 with p.e.g. Ordinates, HA activities, abscissae, concentration of p.e.g. Solid line, HA activities of SNFs after removal of precipitated virus. Interrupted line, differential precipitation curve; dotted line, HA activities of resuspended precipitates. Note the difference of areas included by the interrupted and dotted lines with the abscissae respectively.

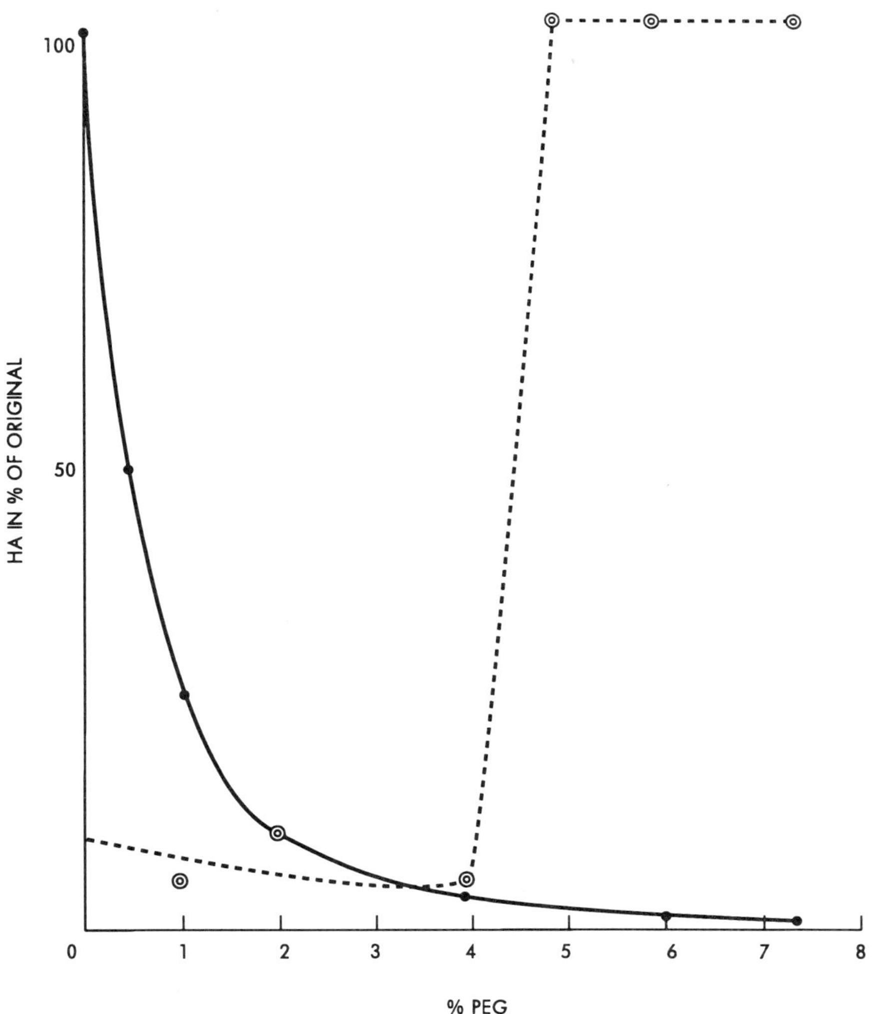

Figure 2. "Aliquot" precipitation of X31 with p.e.g. Data from two batches infected allantoic fluid hence presentation in percentages of original. Solid line, HA activities in SNF after removal of precipitates. Interrupted line, activities of redispersed precipitates.

Table 1. Concentration of X31 Influenza by p.e.g. precipitation and by TL centrifugation*

Materials	HC Activity
Original	640
SNF of 5% precipitate	40
Resuspended 5% precipitate in 500 ml PBS	640
SNF of 6% precipitate	20
Resuspended 6% precipitate in 500 ml PBS	640
SNF of 1% precipitate	40
Resuspended 1% precipitate in 500 ml PBS	40
SNF after think layer centrifugation	20
Resuspended TLR concentrate in 500 ml PBS	640

* It may be stated that in experiments in which smaller volumes of initial infected allantoic fluid were used, similar results were obtained. In these experiments resuspended precipitates obtained with a series of different polymer concentrations were examined in the ultracentrifuge. It was established that p.e.g. concentrations over the range 1 to 4% produced precipitates which were soluble to the extent of less than 5% in the volume of fluid in which they were dispersed for the purpose of ultracentrifugation. No useful purpose would be achieved by presenting their sedimentation diagrams in view of the well defined diagrams presented in Fig. 3.

result given in Table 1 may serve as typical of TL centrifugation of influenza. The virus from 500 ml infected allantoic fluid was concentrated by the TLR in a volume of 3 ml. After redispersion in 500 ml PBS, HA tests showed the solution to have the same titre of the original solution thus indicating complete recovery of virus by TL centrifugation.

Ultracentrifugation

A volume of 2 litres infected allantoic fluid was divided into four equal portions, viz. a,b,c and d.

In portion (a) was dissolved 25 g pulverized p.e.g. The precipitate which formed was centrifuged off at 1000g for 60 min in a refrigerated centrifuge at 4°C. After thorough drainage of p.e.g. solution, the precipitate was suspended in a total of 2.2 ml PBS. This solution was referred to as the 5% precipitated fraction.

In portion (b) was dissolved 30 g pulverized p.e.g. The precipitate which formed was centrifuged off at 1,000 x G for 60 min in a refrigerated centrifuge. After thorough drainage the precipitate was suspended in a total of 2.2 ml PBS. This solution was referred to as the 6% precipitated fraction.

Similarly, portion (c) was treated with 1% p.e.g. and the precipitate suspended in 2.2 ml PBS. This suspension was in turn referred to as the 1% precipitated fraction.

Portion (d) was subjected to centrifugation in the TLR at 6,000 rpm (4000g) for 90 min. The virus concentrate was finally contained in 3 ml. This preparation was referred to as the TLR concentrate or the monomeric concentrate.

The four concentrates were very opaque but they were nevertheless subjected to analytical ultracentrifugation. By removing the light filter and photographing the Schlieren patterns directly with the light from the high pressure mercury vapour lamp, it was possible to obtain sedimentation diagrams. The respective sedimentation diagrams are presented in Fig. 3(a), (b), (c) and (d).

Haemagglutination activities were made on the different concentrates each of which had been resuspended in 500 ml PBS. The results of these assays are presented in Table 1.

Electron Microscopy

Electron micrographs were made of the virus concentrated by TL centrifugation and of that precipitated by 1% p.e.g. Figures 4 and 5 are the respective micrographs. Better photographs could have been obtained if the preparations had been freed of low molecular substances by "washing" but if this were done the true nature of the concentrates would not have been apparent. This would have been the case with the TLR concentration in particular. If the virus concentrate had been "washed", the nucleoprotein and rosettes of haemagglutinin would not have been apparent on the electron micrographs.

It is very clear that precipitation of the virus by 1% p.e.g. formed aggregates in which the individual virus particles are distinguishable with difficulty whereas the virus in the monomeric form obtained by TLR centrifugation lie separate and mostly unassociated on the electron

Figure 3 (a), (b), (c) and (d). Analytical ultracentrifugation diagrams of X31. Conditions of ultracentrifugation similar. Rotor velocity 8 766 rpm; temperature 20°C; medium PBS pH 7.0; exposure intervals 8 min and schlieren angle 60°. Exposures a, b and d taken approximately 40 min after operational velocity had been reached. Exposures cl and 2 taken during acceleration and after 8 min at 8 766 rpm respectively. Photographs taken with light filter removed. Each of the concentrates derived from 500 ml infected allantoic fluid. (a) Resuspended precipitate obtained with 5% p.e.g. in 2.2 ml PBS. Main sedimenting peak that of virus. Peak near meniscus that of protein co-precipitate and negative sedimenting material co-precipitated lipoprotein. Dark region due to intense light scattering of virus suspension. (b) Resuspended precipitate obtained with 6% p.e.g. in 2.2 ml PBS. Interpretation of sedimentation diagram similar to that in (a) above. (c) Resuspended precipitate obtained with 1% p.e.g. in 2.2 ml PBS. Frame (1) taken during acceleration of rotor and (2) after 8 min at 8 766 rpm. The dark region in (1) due to opacity of suspension. This suspension already on base of cell after 8 min. (d) Concentrate by thin layer centrifugation in 3.3 allantoic fluid. Very small protein peak near meniscus and no negative sedimenting component present. Additional component very likely virus capsids formed during prolonged centrifugation in concentration step.

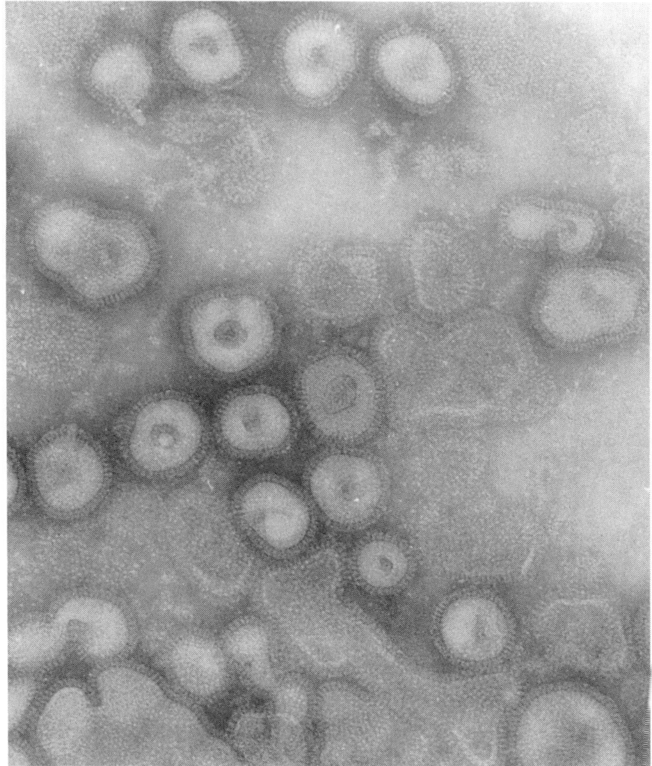

Figure 4. Electron micrograph of the virus concentrated by TL centrifugation. Preparation not freed of allantoic protein. Concentrate similar to that used in analytical ultracentrifuge (Fig. 3d). Note unaggregated state of virus particles.

microscope grid. It was not possible to obtain electron micrographs of the material precipitated with 5 and 6% p.e.g. as the micrographs were obscured by allantoic protein and p.e.g.

There is usually a substantial loss of activity when the virus is spun out of suspension for a second time and washed in PBS. The absence of protective allantoic protein during the step in the purification of the virus very likely allows disintegration of the particles resulting in a loss of HA.

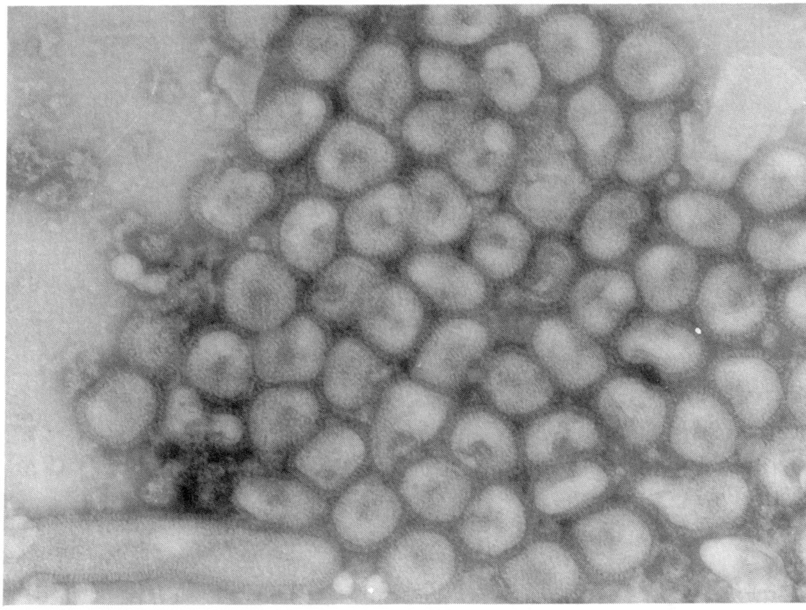

Figure 5. Electron micrograph of virus precipitated with 1% p.e.g. Preparation similar to that in ultracentrifugation experiment (Fig. 3c). Note the aggregated state of the virus particles.

This may be prevented by the addition of 0.1% bovine serum albumin to the wash fluid. In the present context, however, albumin was not added as this would interfere with the protein determination of the concentrate and with electron microscopy. Influenza virus may be spun out of infected allantoic fluid at 6,000 rpm (5000g) for 90 min at 20°C to the extent of approximately 95% or more using the TLR.

Protein assay on X31 Infected Allantoic Fluid

It was found necessary to concentrate infected allantoic fluid 10-fold for protein assay as the protein content of allantoic fluid is too low for accurate determination by the biuret technique. To determine what portion of the total protein present in the allantoic fluid is removed by 4% and 7% p.e.g.,

Table 2. Protein concentration in SNF after removal of the virus by different means from 10 x concentrated infected allantoic fluid

SNF from	g Protein in 100 ml 100 ml
Thin layer ultracentrifugation	0.21
Precipitation with 4% p.e.g.	0.21
Precipitation with 7% p.e.g.	0.16

the concentrated allantoic fluid was divided into three equal portions - (a), (b) and (c). Portion (a) was rendered free of virus by centrifugation in the TL rotor and this served as reference. Portion (b) was treated with 4% p.e.g. and the ensuing virus precipitate centrifuged off. This concentration of polymer removed more than 95% of the virus (see Figs. 1 and 2). Portion (c) was treated with 7% p.e.g. and the bulky precipitate centrifuged off. Protein assays were made on the respective supernatant fluids, the results of which are presented in Table 2.

Protein Assay on Virus Concentrates

As the mass of virus particles in infected allantoic fluid is small in comparison with the allantoic protein, removal of the virus by precipitation with p.e.g. would not influence the protein contents of the allantoic fluid to such an extent that it would be reflected by a reduction of the total protein concentration as assayed by the biuret technique. Therefore, determinations of the protein content of the virus purified by TL centrifugation and of that purified by precipitation with 4% p.e.g. were made. For this purpose 670 ml clarified infected allantoic fluid was divided into two equal portions of 335 ml each. The virus from one portion was concentrated by TL centrifugation to a final volume of 9.5 ml (this required 3 centrifugations at 6,000 rpm for 120 min). This concentrate was diluted to 100 ml in PBS and the virus centrifuged down and contained in 2 ml PBS. The volume was increased to 3 ml prior to protein determination.

The virus from the second portion was precipitated with 4% p.e.g. After removal of extraneous protein by washing with PBS the precipitate was dispersed in 3 ml PBS.

Table 3a. Protein associated with X31 strain of Influenza Virus as purified by TL centrifugation at 6,000 rpm for 2 hours at 20°C. 110 ml infected fluid centrifuged at a time 335 ml total.

MATERIAL	HA ACTIVITY	Total protein mg in 3 ml	REMARKS
Original (335) ml	1280	-	-
SNF after lst TL centrifugation	20	-	-
SNF after final TL centrifugation (wash)	20	-	-
Final virus concentrate in 3 ml PBS	32000	1.98	Suspension blue and transparent

Table 3b. Protein associated with X31 strain of Influenza Virus as purified by precipitation with 4% p.e.g.

MATERIAL	HA ACTIVITY	Total protein mg in 3 ml	REMARKS
Original (335) ml	1280	-	-
SNF after 4% precipitation	10	-	-
SNF after wash of precipitation in 100 ml PBS	640	-	-
Final virus concentrate in 3 ml PBS	5020	3.18	Suspension white and opaque

The following assays were carried out:

(i) Total protein concentration in each of the two concentrates;
(ii) HA activity on the original;
(iii) HA activities on the SNF's after removal of the virus by thin layer centrifugation, and, after precipitation with 4% p.e.g. In addition, HA assays were also made on the two virus concentrates and the two virus "wash" fluids (PBS). The results of this experiment are presented in Table 3.

Antigenicity

The results on the antigenicity experiments in guinea pigs and details of the "vaccines" are presented in Table 4. Details of the virus used for immunizing guinea pigs are presented in Table 5.

DISCUSSION AND CONCLUSION

Influenza virus may be partially precipitated by polyethylene glycol (p.e.g.) m.w. 6,000 at polymer concentrations as low as 1 g/100 ml. At 3 to 4% p.e.g. more than 95% of the virus may be removed from solution without affecting the concentration of the allantoic proteins. The virus precipitate dispersed in a volume of electrolyte equal to that of the original, does not yield the HA titre expected, it may in fact be as low as 15-25% of the original. The dispersed precipitate was apparently insoluble to a large degree as the main portion would be centrifuged from suspension at 500 to 1,000 x G in approximately 10 min. It appeared, therefore, that the virus formed aggregates of low solubility. As this is an important factor in relation to the antigenicity of the virus, further experiments were performed to substantiate this observation.

Electron microscopy indicated a looser association of the virus particles when precipitated with 1 to 4% p.e.g. than that observed with the unassociated or monomeric virus concentrated with the TL rotor. Ultracentrifugation analysis of the monomeric, the 1, 5 and 6% p.e.g. precipitates indicated that the precipitate obtained with 5 and 6% polymer p.e.g. concentration could be redispersed in its monomeric form but that those obtained with 1% p.e.g. were only very sparingly dispersed monomerically, with approximately 95% remaining in the associated form. The virus in this state did not form a schlieren boundary when centrifuged in the analytical ultracentrifuge as it was already centrifuged out of suspension by the time the operational rotor velocity had been reached. Evidence that the virus is present in the precipitate is provided by the protein reactions of the resuspended precipitate after treatment with 4% p.e.g. and that of the pure virus suspension obtained by thin layer centrifugation. Comparing the two protein reactions, it would appear that an estimated 66% of the virus was lost during the lengthy purification procedure with the TLR, the main loss being during the wash in 100 ml PBS as it is well known that influenza virus tends to dissociate in electrolyte solutions in the absence of protecting proteins. The aim was to concentrate the virus 100-fold, but the recovery of the HA activity was approximately 1/3 of that originally present. If 1.98 mg protein (Table 3) is the total

Table 4. HA inhibition activities (1/titre) of guinea pig sera following vaccination with x31 strain of Influenza

G. pig serum	Untreated		RDE treated	
	Monomeric Vaccine			
1	2560		5120	
2	1280		1280	
3	1280		10240	
4	5120	Ar.M.	10240	Ar.M.
5	320	= 1730	1280	= 5630
6	1280		5120	
7	1280		5120	
8	1280		2560	
9	2560		5120	
10	640		10240	
	Aggregated Vaccine			
11	1280		5120	
12	10240		1280	
13	2560		10240	
14	10240	Ar.M.	10240	Ar.M.
15	10240	= 4936	10240	= 7040
16	5120		10240	
17	2560		5120	
18	1280		2560	
19	2560		5120	
20	1280		10240	
	Ether Dissociated Vaccine			
21	2560		10240	
22	640		10240	
23	320		1280	
24	1280	Ar.M.	1280	Ar.M.
25	640	= 888	5120	= 4338
26	320		640	
27	640		2560	
28	320		5120	
29+	1280		2560	
30	-			

+ Died Ar.M. Arithmic mean

Table 5. Details of virus used for immunizing guinea pigs:

VIRUS MATERIAL	HA CONTENT
320 ml allantoic fluid	1,280 - 2,560
30 ml concentrate	20,000
2 ml concentrate from thin layer rotor	100,000 (monomeric vaccine)
2 ml containing 4% p.e.g. precipitate	2,000 (aggregated vaccine)
2 ml containing 4% p.e.g. precipitate	2,000 (ether dissociated vaccine)

Ether treated material not titrated
HAI (haemagglutination inhibition) tests done with 0.5% fowl red cells; test virus 4 HA doses. Serial two-fold dilutions of guinea pig serum in citrate saline pH 7.2. RDE treatment - standard practice

protein associated with approximately 30% of the virus present, the original allantoic fluid contained 6.6 mg viral protein. This amount of protein is to be compared with the 3.18 mg, the mass of the viral protein in the material precipitated by 4% p.e.g.

Protein determinations on the infected allantoic fluid after removal of the virus from suspension by various concentrations of p.e.g. and by thin layer centrifugation showed that only the virus was removed by 4% polymer but that 7% p.e.g. removed the virus as well as approximately 30% of the allantoic protein. This fact provides a simple explanation as to why the virus is not aggregated into partially soluble clusters by 7% (Kanarek et al[2]) and why this happened at 4% p.e.g. At the higher polymer concentration both protein and virus are precipitated and in the "pellet" centrifuged out of suspension. Allantoic protein and virus are thoroughly mixed and consequently the virus particles are separated from each other by protein precipitate. When redissolving the mixed precipitate, the whole mass enters solution and the virus particles are dispersed as they were in their original state. Only the virions are precipitated at 4% and lower concentrations of polymer, and in the precipitate they are forced into closely packed clusters by the displacement properties of the polymer. As a result of the lipoprotein nature of the haemagglutinin on the surfaces of the virions, the particles would tend to adhere mutually through the lipoprotein moieties with complete displacement of the allantoic proteins. These binding forces allow only a minor fraction of the virus precipitate to exist in the monomeric state when the precipitate is redispersed (see ultracentrifugation diagram). Forcing the redispersed precipitate at 500 Kg/cm^2 pressure through a small orifice of the cell disintegrator (Polson[4]) causes liberation of the nucleoproteins from the virions but fails to separate the virus capsids from one another due to intense interlocking forces.

Thermodynamic theory predicts why the clustering of the virus into sedimentable aggregates is a slow process and not instantaneous as is observed when the allantoic protein is precipitated with p.e.g. The escaping tendency from curved surfaces is inversely proportional to the radius of the curved body of the substance. Regarding the solubility as being directly proportional to the escaping tendency or difference in molar free energy, F, the equation (Lewis and Randall[5]):

$$\Delta F = \frac{2rV}{r}$$

in which *r* is the surface tension, V the molar volume and r the radius of the curved surface, should predict the behaviour of the virus clusters in suspension. The entities *r* and V are the same for all the aggregates, therefore the radius r is the only factor which would influence the solubility of the precipitate. If the virus precipitate is composed of clusters of varying radii, it may be expected that the smaller particles would dissolve and be aggregated into the larger units. The transfer of the monomeric virus particles is time dependent as a consequence of their low diffusion rate, and the final state of equilibrium may only be reached after several hours. The virus which failed to be precipitated after treatment with p.e.g. may very likely be the "monomeric" virus in the process of being incorporated in the aggregates and had the reaction been allowed to proceed for a longer time, their numbers in suspension would have been smaller.

Indirect proof for the existence of the aggregates was provided by the use of the virus in this form as an antigen to evoke antibodies in guinea pigs. Using suitable controls, such as the same amounts of the virus in the monomeric and ether dissociated states, the aggregated antigen proved to be antigenically superior. The aggregation of the virus together with its slow release into the system of the infected animal may be regarded as being similar to an adjuvant effect. As a corollary it may be added that influenza virus freed of potentially dangerous allergenic extraneous allantoic protein and aggregated by p.e.g. may well become the most potent "killed" influenza vaccine for the purpose of human immunization. Although not mentioned in the text, the virus inactivated by 0.1% formaldehyde may similarly be aggregated by p.e.g. This will remove doubts which may be raised about whether the inner virus in the clusters are reached by the formaldehyde (Polson, et al[1]).

REFERENCES

1. A. Polson, A. Keen, C. Sinclair-Smith and I.G.S. Furminger. J. Hyg. Camb., *70*, 255-265 (1972).
2. A.D. Kanarek and G.W. Tribe. Nature, London, *214*, 927-928 (1967).
3. A. Polson and L.M. Stannard. Virology, *40*, 781-791 (1970).
4. A. Polson. Prep. Biochem., *3*, 243-256 (1973).
5. G.N. Lewis and M. Randall. Thermodynamics and the Free Energy of Chemical Substances, 252, (1923). McGraw-Hill Book Co., Inc., New York and London.

18. Isolation of a Cytoplasmic Polyhedrosis Virus by Physical and Immunological Techniques*

ABSTRACT

Procedures are described for the separation and purification of cytoplasmic polyhedrosis viruses obtained from multiply infected *Heliothis armigera* larvae. Separation was achieved by differential centrifugation and density gradient zone electrophoresis followed by complexing with nuclear polyhedrosis virus specific antibody. The yield of cytoplasmic polyhedrosis virus was increased by passage in larvae reared on synthetic media.

INTRODUCTION

In order to make a comparative study of the cytoplasmic polyhedrosis virus (CPV) of the bollworm *Heliothis armigera* and the CPV of other hosts, it was necessary to obtain pure CPV virions. As *H. armigera* often has mixed infections of the two polyhedrosis viruses, nuclear and cytoplasmic, isolation of pure CPV from the gut tissue is not easily accomplished. Nuclear polyhedrosis virus (NPV) infection causes the larvae to die within 4-8 days and by this time the cell damage is so extensive that the low concentrations of cytoplasmic polyhedra present are contaminated with nuclear polyhedra.

---

*From: R. Rubinstein, L. Stannard and A. Polson, Prep. Biochem., *5*: 79-80 (1975).

Initial attempts to separate the cytoplasmic from the nuclear polyhedra using the isopycnic centrifugation method failed[2]. Similarly, neither rate zonal nor isopycnic density gradient centrifugation were successful in separating the virions. By using a combination of differential centrifugation, zone electrophoresis and complexing the NPV with antibody, virions of the two types of polyhedra were successfully separated. The preparative procedures for isolating pure CPV will be described in detail.

MATERIAL AND METHODS

Diseased larvae (obtained from the Department of Agricultural Technical Services, Division Pest Control, Mowbray, Cape) were macerated in a mortar and suspended in distilled water. The resulting mass was filtered through four layers of cheese-cloth and the filtrate centrifuged at 6,000 x g for 10 min in a Spinco 40 rotor to sediment the polyhedra. Contaminating bacteria and debris were removed by suspending the pellet in distilled water and shaking with an equal volume of arcton 113 (Imperial Chemical Industries). The aqueous phase containing the polyhedra was separated and collected. This procedure was repeated several times and the combined aqueous fractions were centrifuged at 6,000 x g for 10 min. The pellet of polyhedra was washed once in distilled water and treated with a mild alkali solution containing 0.02 M Na_2CO_3 and 0.05 M NaCl, pH 10.4, to release the virions. An initial digestion period of 20 min at room temperature was followed by centrifugation at 6,000 x g for 10 min and the sediment of residual inclusion bodies treated with the same alkali solution for a further 30 min. The combined supernatant fluids (SNF), which contained the released virions, were diluted with distilled water to prevent further digestion. This was the starting material for purification of the CPV in all experiments. The following methods were applied:

Differential Centrifugation

In order to ascertain the conditions for the optimal separation by differential centrifugation it was essential first to determine the sedimentation coefficient of the viruses. The mixture of virions was concentrated by centrifugation and the pellet resuspended in 0.8 ml normal saline. Sedimentation coefficients were determined using the Spinco Model E analytical ultracentrifuge. A series of sedimentation diagrams are presented in Fig. 1a. On the basis of the established sedimentation coefficients, three repeated centrifugation at 30,000 x g for 13 min in the

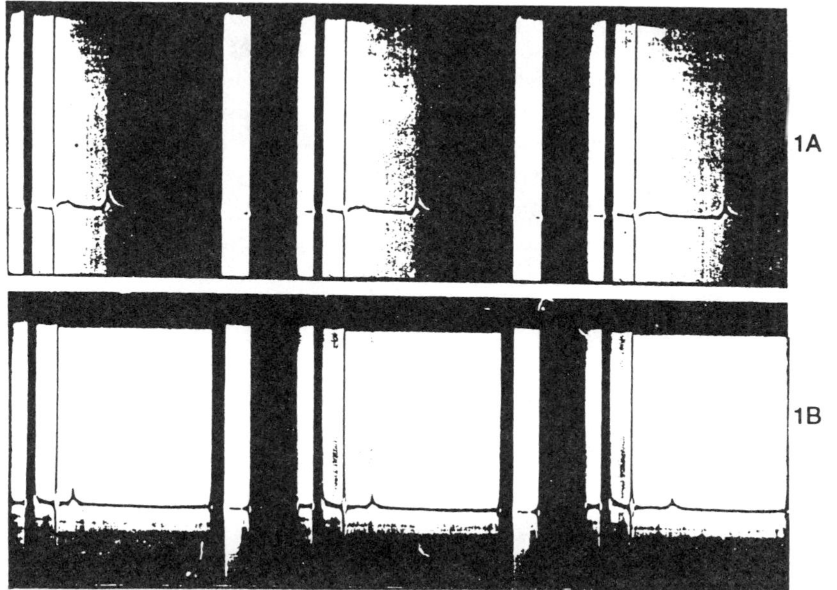

Figure 1a. Analytical sedimentation pattern of "CPV" (Minor peak) and "NPV" (main peak) virus particles. Sedimentation is from left to right. The rotor velocity was 8766 rpm at a temp of 20°C and the exposures taken at 4 min intervals. The Schlieren angle was 60° and the medium was 0.14 M sodium chloride.

Figure 1b. Analytical sedimentation pattern of pure CPV particles. Sedimentation is from left to right. The rotor velocity was 8766 rpm at a temp of 20°C and exposure taken at 8 min intervals. The Schlieren angle was 50° and the medium was 0.14 M sodium chloride.

SW 39 rotor removed the bulk of the NPV particles. The final supernatant was centrifuged at the same speed for 40 min to sediment the CPV.

Density Gradient Zone Electrophoresis

Zone electrophoresis was performed using the apparatus described by Polson and Cramer[3] with buffer and sucrose gradient as used by Polson and Russell[4]. Briefly the apparatus consists of a vertical U-tube with an electrode vessel at the top of each limb. One limb in which the electrophoresis is conducted contains a logarithmic sucrose density gradient. The operational region of the gradient is approximately linear. The partially purified virus sample derived from differential centrifugation was introduced into the column together with phenol red as reference substance. A voltage gradient of 3.5 v/cm was applied for 18-20 hr. On completion of electrophoresis, measurements were made from the origin to the centres of the virus and the phenol red zones. The ratio of the respective distances of migration is termed the RØ value[6]. When the phenol red had migrated approximately 16-18 cms, aliquots representing 1 cm of column length were collected and virus was recovered from the diluted aliquots by centrifugation at 30,000 x g for 40 min. The position of the virus was determined by electron microscopy. This procedure was essential for CPV where no zone of light scattering was visible.

Treatment with Specific NPV Antibody

Antiserum to NPV was prepared by injecting rabbits intramuscularly with a mixture of NPV (purified from single infections by gradient centrifugation) and Freund's adjuvant three times at weekly intervals. Virus suspended in phosphate buffered saline (PBS), pH 7.0, was mixed with an equal volume of the NPV antiserum diluted 1/10, incubated at 37°C for 30 min and left at 4°C overnight. The sample was diluted with PBS, centrifuged at 17,000 x g for 15 min in a SW 39 rotor to remove NPV complexed with antibody and the supernatant was recentrifuged at 30,000 x g for 40 min to sediment the CPV particles.

The virus thus obtained was used for infectivity experiments with 6-day old *H. armigera* larvae reared at 30°C on synthetic medium[1]. Surface sterilization of the eggs with dilute sodium hypochlorite was used to reduce possible transmitted virus in the larvae. Six to eight days after ingestion of infected medium the larvae were killed and their midguts removed. The midguts and the remainder of the larval tissue were separately extracted and examined for virus content.

Electron Microscopy

The negative staining technique was used during examination of the purity of virus samples at all stages. Virus pellets were suspended in a drop of distilled water mixed with an equal quantity of 2% phosphotungstic acid, placed on a formvar-carbon-coated grid and examined in a Siemens Elmiskop 1A.

Figure 2. Electron micrograph showing the proportion of CPV and NPV virions found in an extract of naturally infected larvae of *H. armigera*.

RESULTS

A series of sedimentation diagrams obtained in the analytical ultracentrifuge with a natural mixture of NPV and CPV is shown in Fig. 1A. The final purified virus sample appears in the sedimentation diagram, Fig. 1b.

The sedimentation coefficients calculated were 389 and 1188 S for the minor and main peaks representing CPV and NPV respectively. Using these values together with the known minimum time required to sediment

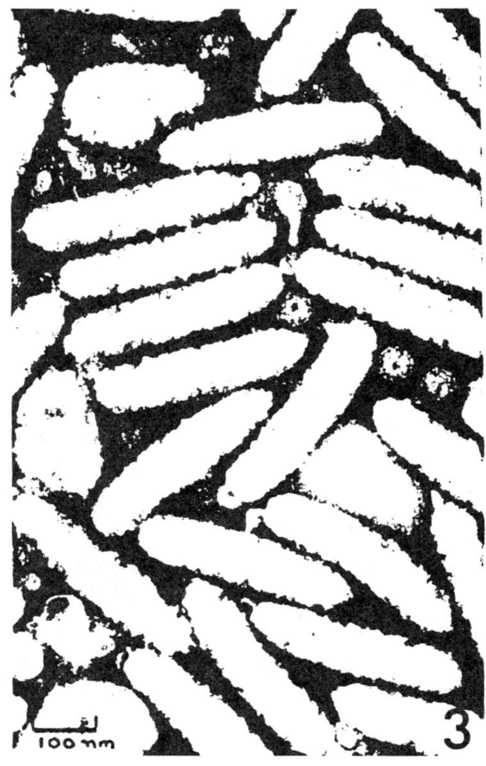

Figure 3. The resuspended pellet after 30,000 x g for 13 min, predominantly NPV particles.

Burnupena cincta haemocyanin (S_{W20} = 100) it was possible to calculate the most effective period for ultracentrifugation in the preparative rotor to remove most of the NPV with minimal loss of CPV from the suspension. This period was calculated to be 13 min at 30,000 x g. Electronmicrographs of the virus contained in the initial mixture, in the pellet and SNF under these conditions of centrifugation are presented in Figs. 2-4. It was found that further centrifugation resulted in damage to CPV particles. The majority of the contaminating complete NPV virions and "empty" NPV nucleocapsids remaining in the CPV fraction after

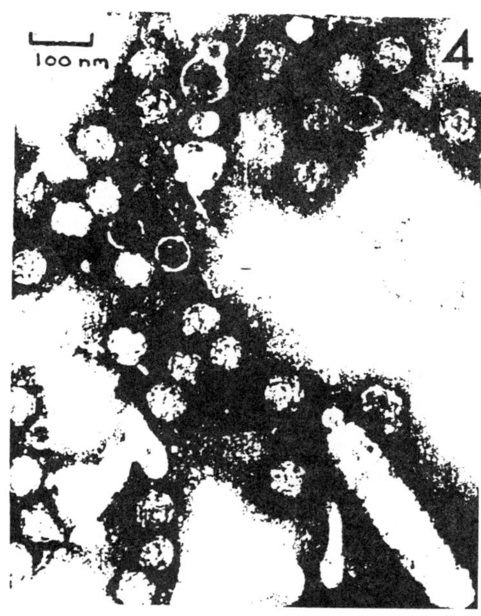

Figure 4. The SNF after ultracentrifugation at 30,000 x g for 13 min showing a predominance of CPV.

Table 1.

Virus particle	Relative electrophoretic mobilities in RØ values
(i) CPV virions released from polyhedra	1.0
(ii) Enveloped NPV virions	0.59
(iii) "Empty" NPV nucleocapsids	0.82

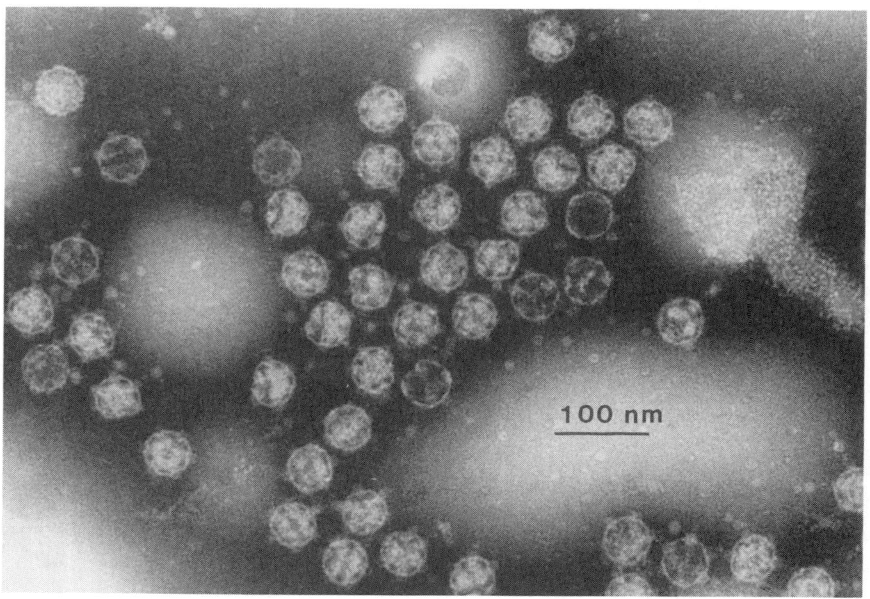

Figure 5. The CPV virions isolated from the midgut of *H. armigera* larvae which were infected with the final CPV sample treated with NPV antiserum.

differential centrifugation were removed by density gradient zone electrophoresis.

Table 1 indicates the relative electrophoretic mobilities of the different virus particles expressed as RØ values. The CPV virions with a calculated RØ of 1.0, migrate with the phenol red and were therefore not visible as a discrete band. The majority of complete NPV particles were well separated from the faster migrating CPV in the density gradient, but the electrophoretic heterogeneity of the NPV nucleocapsids resulted in some contamination of the CPV zones. These contaminating particles were removed by complexing with specific NPV antibody followed by centrifugation at 12,000 x g for 10 min.

Assessment of the purity of the final CPV suspension was made by infecting virus-free larvae with the material obtained after antibody treatment and examination of the purified tissue extracts by electron microscopy.

Only CPV virions, both free and occluded, were found in the electron micrographs of extracts of the midgut (Fig. 5). No CPV particles were found in other tissue of these larvae, or in further passage of this CPV stock. Only a single sedimenting peak was found on analysis in the analytical ultracentrifuge (Fig. 1b). A sedimentation coefficient of 389 S was calculated. This confirmed the assumption made above that the minor sedimenting peak in Fig. 1a represents the CPV virions.

DISCUSSION

Multiple infection by CPV and NPV has been described in the silkworm, *Bombyx mori*[5]. In this laboratory such mixtures have frequently been encountered in *Heliothis armigera* but attempts to separate polyhedra of these viruses using isopycnic centrifugation proved unsuccessful presumably due to size range of polyhedra for both viruses. However, CPV was successfully isolated and purified from a mixture of released CPV and NPV virions.

Differential centrifugation was the most convenient preliminary isolation step but complete separation by this method could not be achieved. This was probably a result of retardation of particles of the size of NPV on the wall of the tubes in which the mixture was ultracentrifuged. Damage to CPV particles limited the use of repeated centrifugation.

When differential centrifugation was followed by density gradient zone electrophoresis, further purification of CPV resulted, but occasional NPV nucleocapsids remained as contaminants. CPV and enveloped NPV virions differed sufficiently in their electrophoretic mobilities to be widely separated by this technique. The findings that some NPV nucleocapsids migrate at a rate similar to that of the CPV virions may be due to loss of surface protein and the reorienting of charged groups resulting from the release of the nucleoprotein.

After differential centrifugation and zone electrophoresis, the CPV virus was sufficiently pure for most biophysical and biochemical studies. For infectivity experiments and the production of antiserum, a pure sample was required which necessitated the additional treatment with NPV antiserum. The supernatant material, after removal of the NPV-antibody complexes and subcultured in larvae, resulted in an infection from which only CPV could be harvested.

These procedures provide a means of achieving effective separation of CPV from NPV while retaining the infectivity of the particles, and may find a useful application in the isolation of other viruses.

ACKNOWLEDGMENTS

We would like to thank Professor A. Kipps for his interest in the work and the Department of Agricultural Technical Services for financial aid and the Division of Pest Control, Mowbray, for the original diseased larvae.

REFERENCES

1. J. Bot, S.A.J. Agri. Sc. *9*, 535-538 (1966).
2. M.E. Martinogi, J. Virol., *1*, 646-647 (1967).
3. A. Polson and R. Cramer, Biochim. Biophys. Acta, *29*, 187-192 (1958).
4. A. Polson and B. Russell, in "Methods in Virology", Vol. II, Ed. K. Maramorosch and H. Koprowski, Academic Press, New York, London, 1967, pp. 391-426.
5. K.M. Smith and N. Xeros, Parasitology, *43*, 178-185 (1953).
6. M.H.V. van Regenmortel, S.A. Med. J., *42*, 118 (1968).

19. Purification of Filamentous Plant Viruses by Thin Layer Centrifugation (Applied to TMV, SCMV, PVX, SCV and YMC Viruses)*

ABSTRACT

The construction of a modified thin layer ultracentrifuge rotor is described. This rotor was used in the purification of five filamentous plant viruses, viz. TMV, SCMV, PVX, SCV and YMC. The purification and concentration of these viruses in their monomeric forms is hazardous when conventional "tube" rotors are used since they invariably result in dissociation and aggregation of the virus particles. Using the thin layer rotor these infective agents may be concentrated in volumes of fluid equal to approximately 1% of the starting suspension and not as pellets obtained after ultracentrifugation in conventional "tube" rotors. Electron microscopy revealed that the virus particles concentrated by thin layer centrifugation were not aggregated and that only few fragments of the virus filaments were present in the final preparations.

INTRODUCTION

Among the major factors detrimental to filamentous viruses during their isolation from infected plant tissue are:

---

*From: A. Polson and A. Kiefer, Prep. Biochem. *5*:199–210 (1975).

(a) formation of polyphenolic compounds during initial extraction of the plant pathogen[1]. These react irreversibly with plant proteins as well as with the virus present in the extract;
(b) attachment of extraneous plant host proteins to the virus filaments;
(c) association of virus particles into paracrystalline clusters which cannot be dissociated into their monomeric form without causing extensive rupture of the filaments;
(d) breakage of virus filaments into shorter lengths resulting from the application of harmful physical forces during isolation and purification.

The formation of polyphenolic compounds is caused by the presence of polyphenoloxidaze. This enzyme may be inhibited by the addition of sodium diethyl dithiocarbamate[1].

As a precautionary measure against other unknown oxidation agents, 2 mercaptoethanol and sodium thioglycolate may be added during preliminary disintegration and extraction of the plant material.

Attachment of extraneous plant proteins to viruses occurs in many instances through "calcium bridges" between virus and proteins. The paracrystalline clusters of filamentous virus particles may result from calcium bridges, or from associations such as virus-protein-calcium-protein-virus. To render the calcium or other responsible divalent ions non-ionic or unreactive they may be complexed with known chelating agents. By adding these divalent ion binding substances to the strongly reducing mixture during the initial disintegration of the plant tissue, aggregation resulting from this cause may be effectively eliminated. McIlvane's citrate buffer in the pH range 7.0 - 8.0 to which EDTA had been added serves the purpose well.

Fragmentation of virus filaments may be the result of shearing forces during the initial disintegration of the infected plant material. The use of a high speed Waring blender may be very damaging to the virus filaments. Effective disintegration of the plant tissue may be accomplished by crushing the infected plant material material in a meat mincer in the presence of a solution of the mixture of reducing agents, enzyme inhibitor and metal ion complexing substances.

The major cause of aggregation in parallel is close packing into pellets under excessive centrifugal stress and the main cause of breakage of the virus filaments is redispersion of the pellets by trituration. If the virus particles could be concentrated into a small volume of fluid by a lesser centrifugal force than that normally used for the purpose in conventional preparative ultracentrifugation, unassociated monomeric particles could be obtained. With this object in mind a centrifuge rotor was designed and constructed in which intact filamentous viruses could be concentrated into a volume of fluid equal to approximately 1% of that subjected to centrifugation.

The new rotor was designed to fit the driving shaft of the Spinco preparative centrifuge, and is referred to as the 120 thin layer rotor (120 tlr).

MATERIALS AND METHODS

The Thin Layer Rotor

In previous communications[1,2,4] centrifuge rotors were described in which the distance of sedimentation was greatly reduced by the use of conically slanting "baffles" in a circular cavity. The baffles were machined from a suitable plastic material since metal would have too high a density differential between it and the aqueous medium in which they are contained. The rotor described here differs from those previously reported, the main differences will be evident from a study of Figs. 1 and 2. The clarifying section containing the two valves is free and may be removed from the rotor for cleaning when necessary. The lid screws on to the periphery whereas in the previous model it was bolted on by six steel screws. The main cavity contains a set of three baffles (Fig. 1b) machined from nylon and unattached to one another. This facilities cleaning and sterilization.

For construction of the baffles nylon was selected in preference to metal, such as aluminum or aluminum alloy, because the density differential between nylon and the aqueous solution is of the order of 0.1 for nylon as opposed to 2.0 g/cc for aluminum alloy, thus reducing the danger of damage to the baffles during centrifugation. In the previous models two baffles were used and these were separated from one another by blocks of plastic material attached to them to form a unit.

An important difference between the earlier and the present model is that of capacity. Whereas previously, 40 ml of suspension could be treated in one operation, the latest model allows a maximum of 120 ml to be centrifuged.

Two valves are inserted into the lid for the purpose of allowing air to enter into the central cavity after centrifugation. The valves also facilitate removal of the supernatant fluid after centrifugation and, if desired, the introduction of further batches into the rotor without removing the rotor's lid.

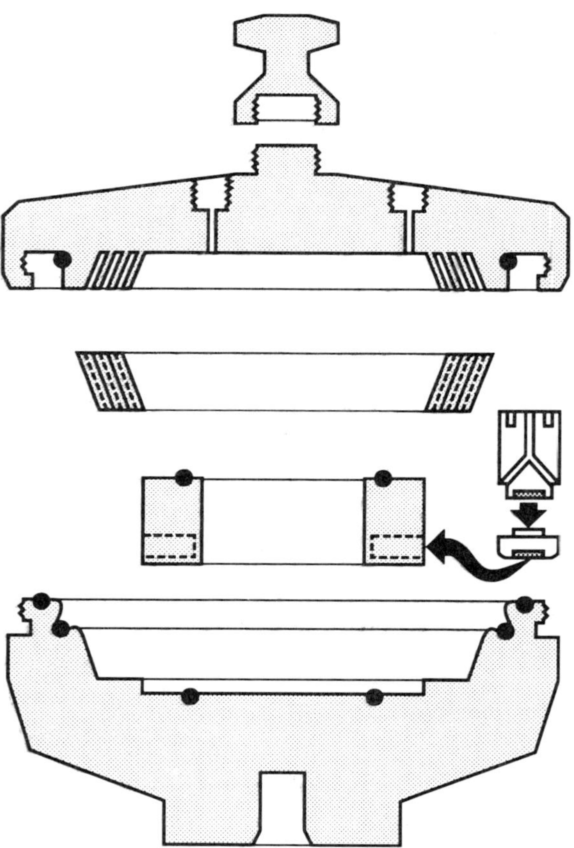

Figure 1a. Cross-section of the thin layer rotor. At the top of the diagram the knob of the lid is shown. Then follows the lid with its three baffles sloping inward at an angle of 112.5° to the surface, its two pressure releasing valves and the O-ring. Below the lid is the assembly of nylon baffles shown in detail in Fig. 1b. The clarifying cavity then follows with the spaces into which the valves fit. Details of a valve may be seen on the right side of the clarifying cavity consisting of two nylon pieces on to which silicon and synthetic rubber pieces are attached, respectively. When in position the synthetic rubber piece closes a tiny hole in the clarifying cavity. At preselected velocities the silicon rubber is compressed and the fluid in the clarifying chamber is allowed to escape through the channel in that part of the valve holding the silicon rubber piece. The velocity at which the valve opens may be controlled by adjusting the tension on the rubber pieces. The main body of the rotor is below the clarifying cavity. The position of the O-rings are shown and that of the serrated polyurethane cord in the receiving cavity. When assembled with the lid screwed on, clearance between the baffles in the lid and those in the main cavity is approximately 1 mm. Diameter of lid 256 mm.

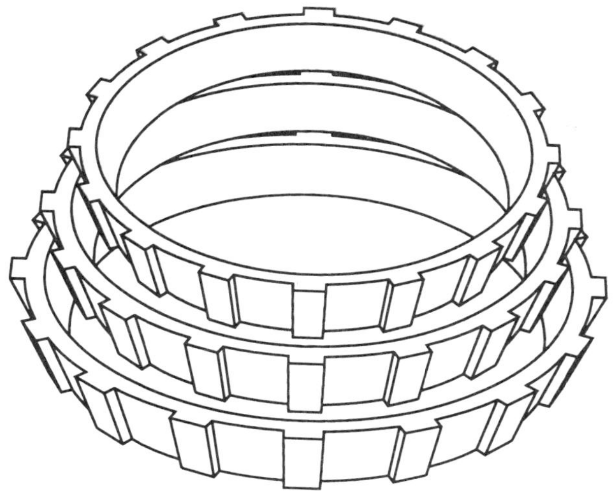

Figure 1b. Set of three nylon baffles showing the spacer ridges (exploded view).

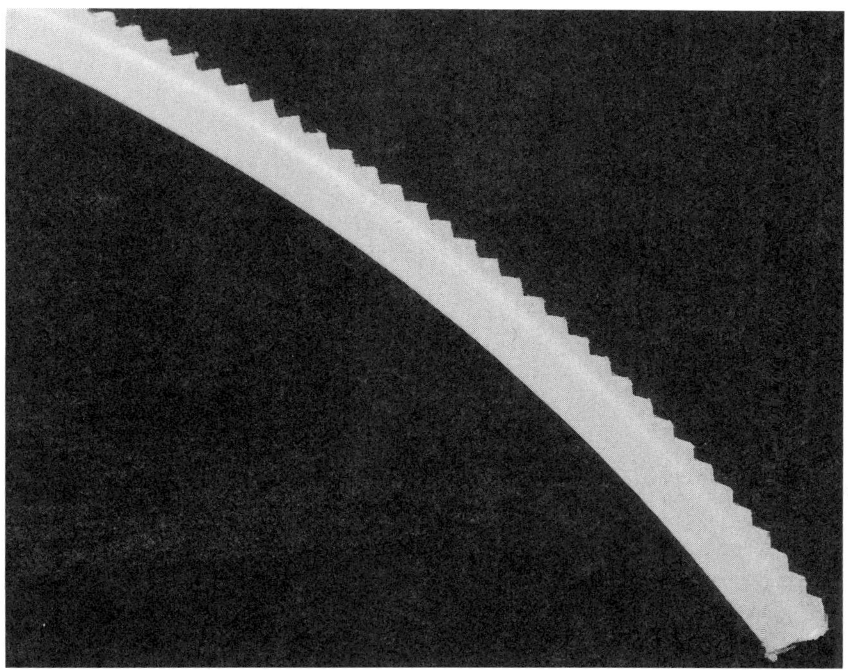

Figure 1c. Photograph (enlarged) of a piece of serrated cord. When in place in the receiving cavity, the virus particles collect as a suspension in the cavities formed by the serrations and the receiving cavity wall.

Figure 2. A photograph of a rotor made of "giduminium" with the lid unattached. The positions of the clarifying chamber, baffles and serrated cord are obvious. This rotor has the same dimensions as the "nylon" version.

Standardization

To avoid excessive centrifugation of unstable materials such as certain arboviruses, it is advisable to sediment such material for a period of time adequate to concentrate 99% of the virus behind the serrated cord in the receiving cavity of the rotor. From a knowledge of the size of the virus an approximate sedimentation coefficient may be anticipated so that the minimum time and rotor velocity may be estimated. A useful reference substance for this purpose is the haemocyanin of the whelk, *Burnupena cincta*. This respiratory pigment has a sedimentation coefficient of 100 Svedberg units[5] and when centrifuged in the present rotor at 15,000 rpm at 7°C the reduction of haemocyanin in the supernatant fluid (SNF) was found

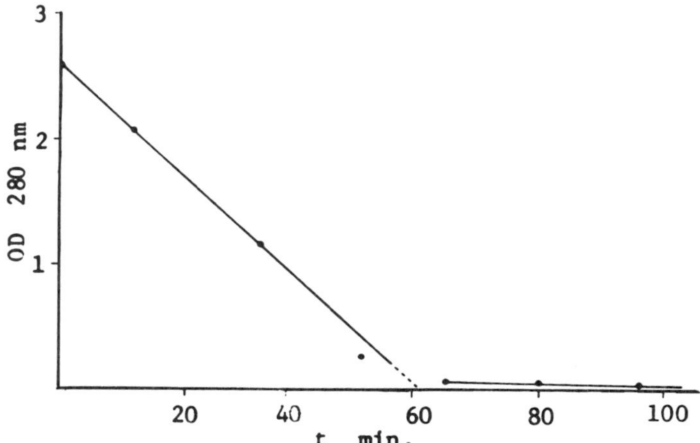

Figure 3. Diagram showing the concentration in terms of u.v. absorption of *Burnupena cincta* haemocyanin of S = 100 Svedberg units after centrifugation for different periods at 15,000 rmp in saline at 7°C in the new 120 t.l. rotor. This diagram is the reference curve (see text).

to be a linear function of time (Fig. 3) up to the point of 98% of its removal. This occurred approximately 60 min after the rotor had reached its operational velocity.

If it is assumed that virus particles would behave similarly, the following equation may be applicable:

$$S_1 t_1 \,(\mathrm{rpm})_1^2 = S_2 \, t_2 \,(\mathrm{rpm})_2^2 \qquad (1)$$

where S_1, t_1, and rpm are the sedimentation coefficient, time of centrifugation and rotor velocity respectively for the reference substance and S_2, t_2 and rpm_2 for the virus to be concentrated.

Since 98% of the haemocyanin is removed from solution in this new rotor in 60 min at 15,000 rpm instead of 30,000 rpm for 90 min in the No. 40 Spinco rotor, the centrifugal force time is reduced by 83.3 percent.

The rotor may also be used for the estimation of approximate sedimentation coefficients using equation (1) and the data for a reference substance of known sedimentation coefficient. Furthermore on account of the relatively low rotor velocities by which plant viruses and other infectious agents may be concentrated, plastic materials of high tensile strength may be used in the construction of a rotor. Nylon appeared to be suitable for the purpose provided that the rotor velocity does not exceed an estimated velocity of 20,000 rpm. Nylon was therefore used in the construction of a 120 tl rotor but the rotor served the purpose on which the construction of the "Giduminium" rotor was based. The nylon rotor was nevertheless spun at 15,000 rpm (15,000 g) and was found suitable for centrifuging viruses into concentrated suspensions. Harmful pelleting of viruses could thus be avoided. Although the nylon rotor did not stretch irreversibly after several operations at 15,000 rpm (25,000 g) its application in virology was abandoned in favor of the "Giduminium" rotor; the reason being that when higher rotor velocities than 20,000 rpm are required the nylon would be unsafe to use. With the rotors depicted in Figs. 1 and 2 approximately all the virus in the fluid may be concentrated 100-fold behind the serrated cord. The depth of the serrations in the plastic cord occupying the receiving cavity determine the volume and therefore the final concentration achieved.

As a precautionary measure against corrosion the metal rotor interior is rinsed with a solution of silicon grease in chloroform after thorough washing with a detergent soap.

Sterilization

Formaldehyde at a concentration of 10% followed by a wash with 70% ethanol, then exposure to ultraviolet light proved to be quite effective for sterilization. Sodium hypochlorite is not suitable.

Preservation of the Sealing Rings

The O-rings play a vital role in the rotor and should be kept in good condition. To ensure against damage which may result in leakage it is essential that their surfaces be coated periodically with silicon grease.

The Viruses

The following 5 viruses were studied: tobacco mosaic virus (TMV), potato virus X (PVX), apple stem grooving virus (SGV)[6], yellow mosaic virus from a *Rubus* species (YMV)[7] and sugar cane mosaic virus (SCMV). The extraction, purification and concentration of TMV, PVX, SGV and

YMV were identical. Infected leaves (50-100g) were ground up in a meat grinder in the presence of 100 ml of a mixture containing 0.2 M Na_2HPO_4 - citric acid buffer pH 7.5, 0.1 EDTA (disodium salt), 0.02 M 2 mercaptoethanol, 0.2 M sodium thioglycolate and 1% Na diethyldithiocarbamate. The plant sap was filtered through two layers of cheesecloth and clarified by centrifugation at 12,000 x g for 10 min. The clear supernatant fluid (SNF) was shaken for 1 min in the presence of an equal volume of analytical grade chloroform. Re-centrifugation at 12,000 x g for 10 min gave a SNF of 100 to 120 ml, adequate for a single charge of the t.1. rotor. The virus suspension was placed in the main cavity of the rotor and spun for 60 min. at 12,000 rpm (20,000g). The virus concentrate (1-1.5 ml) in the receiving cavity behind the serrated polyurethane cord was diluted in 100 ml phosphate buffer of 0.005 M and pH 7.0 and reconcentrated into 1-1.5 ml by centrifugation in the t.1. rotor under the same conditions. A third cycle of centrifugation usually proved adequate for ensuring a monomeric virus suspension significantly free of extraneous materials. Sugar cane mosaic virus was liberated from infected malze leaves by homogenization in McIlvane buffer (0.2 M Na_2HPO_4 adjusted with citric acid to pH 7.0) containing 0.3% 2-mercaptoethanol. The pulp was filtered through two layers of cheesecloth and the sap clarified by centrifugation at 12,000g for 10 min. After dialysis against distilled water for 24 hr the suspension was concentrated by pervaporation through cellophane tubing. The virus precipitated after the volume had decreased to 1/5 of the original. The precipitate was redissolved in 100 ml McIlvane buffer of 0.005 M Na_2HPO_4 at pH 7.0 containing 0.01 M EDTA and subjected to one cycle of t.1. centrifugation at 20,000g for 60 min.

RESULTS

The results obtained with the different filamentous viruses are presented as electron micrographs (Figs. 4 to 9). Figure 4 is that of TMV centrifuged into a pellet in a tube rotor at 108,000 x g for 90 min and left for 16 hr under phosphate buffer of pH 7.0 and molarity 0.05 before being dispersed by gentle trituration. This material served to indicate the damage normally done during purification of filamentous viruses by pelleting followed by redispersion. Fig. 5 is that of TMV concentrated in the t.l. rotor at 18,000g for 60 min. The legends are self-explanatory.

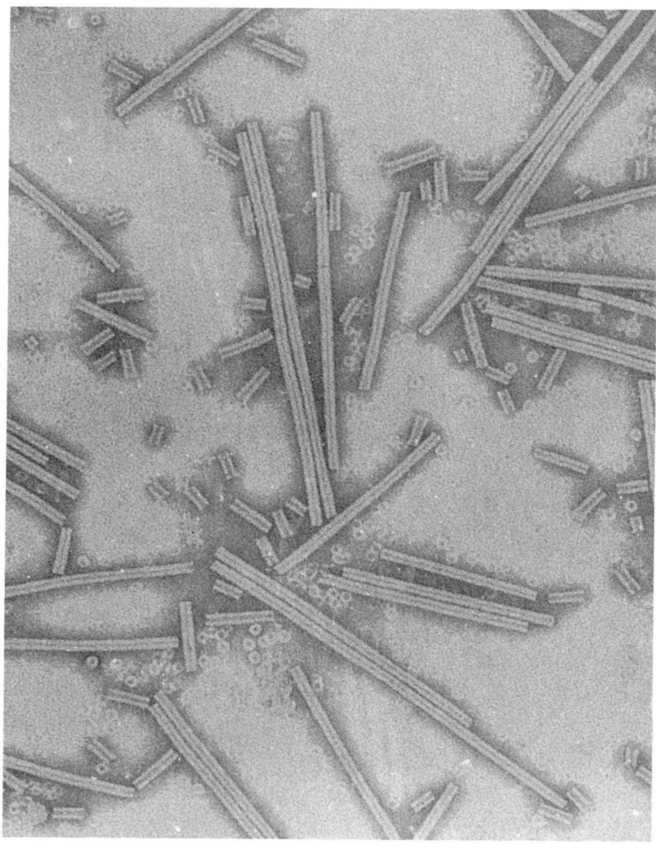

Figure 4. Electron micrograph of TMV centrifuged into and dispersed from a pellet by trituration following conventional preparative ultra-centrifugation. PTA staining as above. Note the large number of virus fragments.

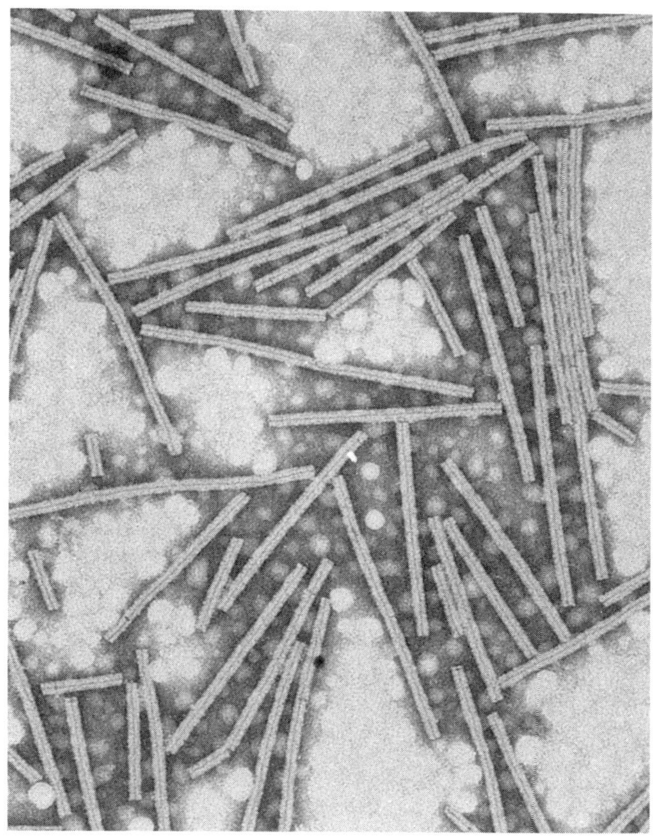

Figure 5. Electron micrograph of TMV virus concentrated 3 x in the 120 t.l. rotor at 20,000 x g for 90 min. The preparation was negatively stained with pta at pH 6.0. Note the small number of virus fragments.

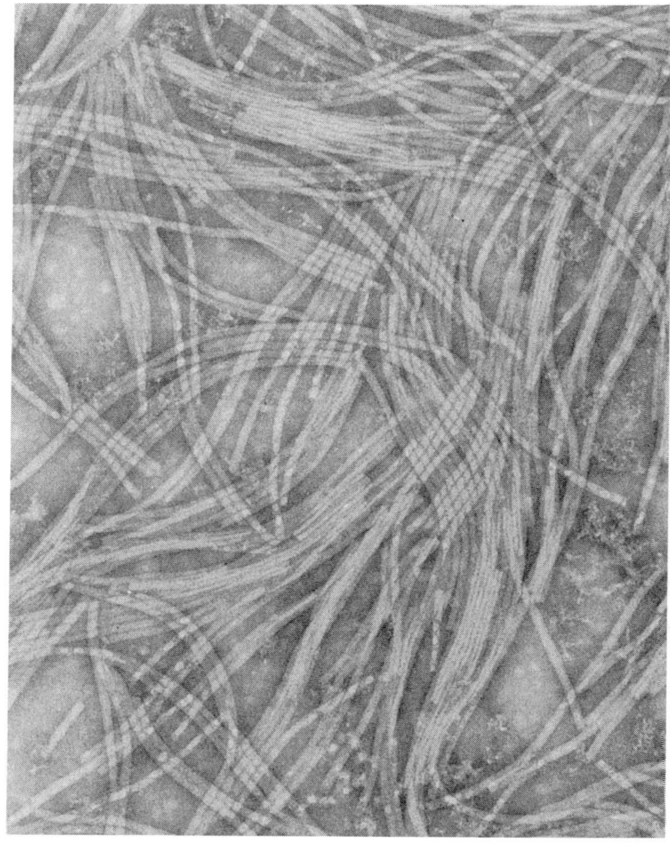

Figure 6. Electron micrograph of PVX concentrated 3 x in the 120 t.l. rotor at 20,000 x g for 90 min. Negative staining as with TMV. Note the unaggregated state of the virus filaments.

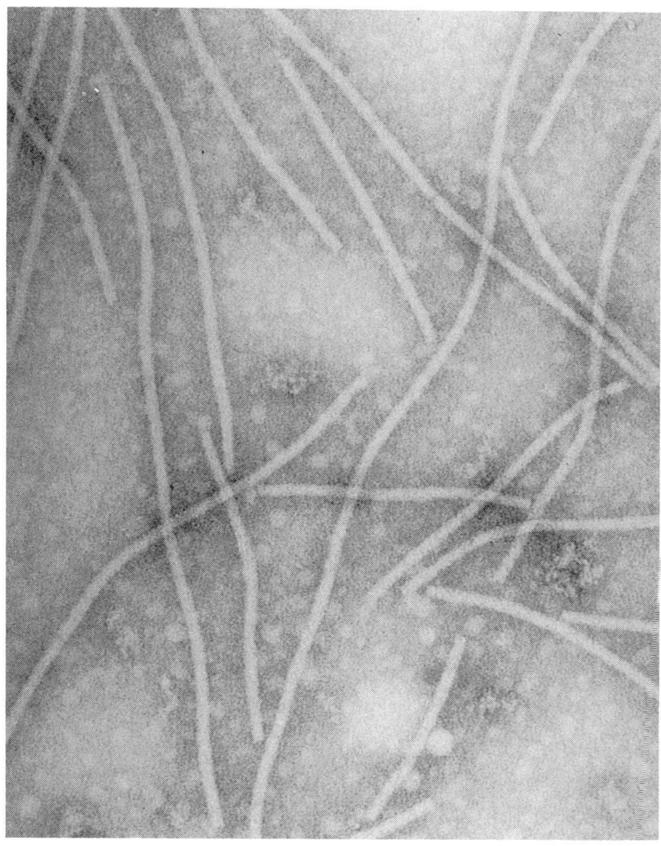

Figure 7. Electron micrograph of SGV concentrated 3 x in the 120 t.l. rotor at 20,000 x g for 90 min. Negatively stained as with TMV.

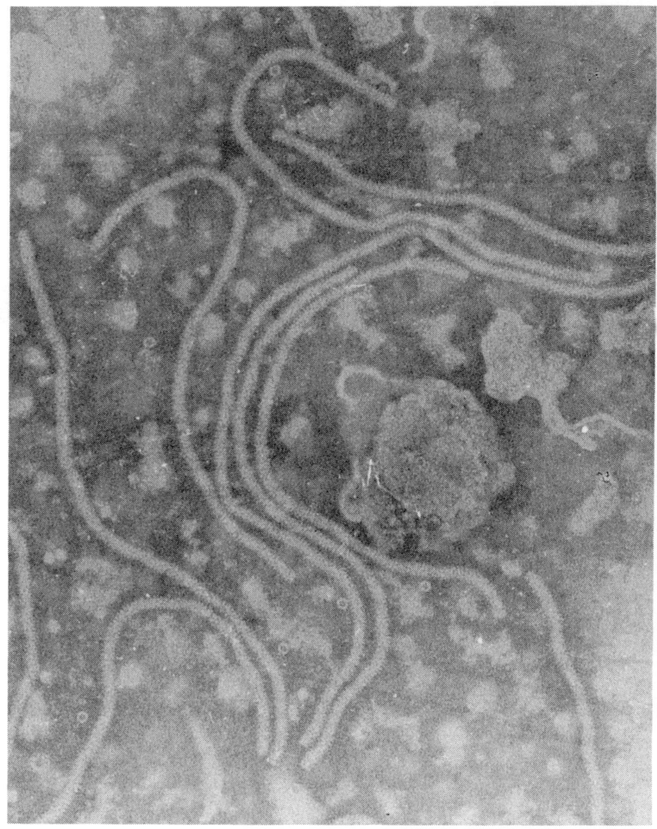

Figure 8. Electron micrograph of YMV concentrated 3 x in the 120 t.l. rotor at 20,000 x g for 90 min. Negatively stained as with TMV.

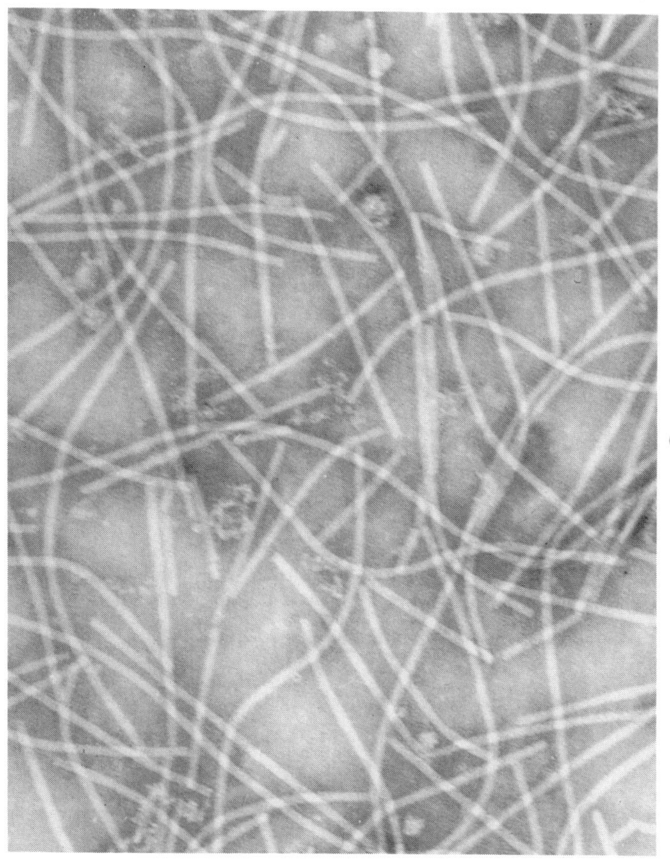

Figure 9. Electron micrograph of SCM concentrated 1 x in the 120 t.l. rotor at 20,000 x g for 90 min. Negatively stained as with TMV.

DISCUSSION

An improved thin layer ultracentrifuge rotor (t.l. rotor) is described. This rotor has a greater diameter and a larger capacity than the model described previously[2,3,4]. Whereas the first model had a capacity of 40 ml the new t.l. rotor can now accommodate up to 120 ml. The rotor is made of an aluminum alloy known in the trade as giduminium which has a tensile strength similar to that of duraluminum. The new 120 t.l.r. is provided with a removable clarifying cavity and a lid which is screwed on to the periphery of the main body of the rotor. The thin layer effect is improved by the addition of an extra baffle. The baffles, made of nylon, do not form an undetachable unit as in the first model but are inserted into the rotor individually. A serrated polyurethane cord is used in place of the serrated nylon cord, because the nylon cord is rigid and tends to spring out from the receiving cavity. This effect may be hazardous as infected material may be scattered. The polyurethane is completely flexible and may be easily removed from the cavity with the aid of two pairs of forceps. If quantities of infectious material greater than 120 ml need to be centrifuged, the virus may first be concentrated by pervaporation or by precipitation with polyethylene glycol and the precipitate dissolved in a suitable volume of buffer.

An important advantage over the conventional "tube" rotor is that a substance may be centrifuged from the supernatant with a quarter of the centrifugal force required by the tube rotor of the same capacity during the same period of centrifugation. In addition, the component centrifuged from suspension into the receiving cavity remains in suspension, not as a pellet which would require trituration for redispersion.

The use of the new 120 version of the t.l. rotor has demonstrated the partial purification of several filamentous viruses, TMV, PVX, SGV*, YMV** and SCMV*** in an essentially undegraded and unassociated form. Two of the viruses investigated, i.e. TMV and PVX are notorious for their readiness to break up into shorter fragments and also to form paracrystalline clusters when subjected to conditions normally used during their isolation by conventional preparative ultracentrifugation. TMV contaminant in PVY° seeding material was removed with anti TMV serum.

The object of this work was to illustrate the use of the new 120 version of the t.l. rotor for preservation of the morphology of filamentous viruses during purification procedures. Complete removal of extraneous materials could be accomplished by subjecting the virus concentrate after t.l. centrifugation to zone electrophoresis in a sugar concentration gradient.

In the present experiments the clarifying chamber of the rotor was not used because the procedures followed for the removal of coarse extraneous matter proved to be adequate for purification and concentration of the filamentous viruses by t.l. centrifugation.

* Apple stem grooving virus was transmitted to and isolated from infected *Chenopodium quinoa* (Willd) leaves. The isolation and characterization of this virus was in turn described in the M.Sc. thesis of W. J. K. van der Walt (1973), Department of Microbiology, University of Stellenbosch. An account of this agent is given by van der Walt and Engelbrecht[6].
** The virus of the wild bramble was described and characterized by Engelbrecht and van der Walt[7]. These authors transmitted the virus to *Chenopodium quinoa* from which they isolated the infective agent in an amount adequate for their purpose.
*** Sugar cane mosaic virus was extracted from infected maize leaves by Mr. D. de Wit and its isolation, purification and characterization formed part of the requirement for B.Sc. (Hons.), 1974, in Microbiology, University of Cape Town.

ACKNOWLEDGEMENTS

The authors wish to express their gratitude for the interest that the Director of the M.R.C. Virus Research Unit, Professor A. Kipps, has taken in the development of the rotor, to the Department of Mechanical Engineering, University of Cape Town, to Dr. S. Eggers of the S.A. Inventions Development Corporation and to the Department of Agricultural Technical Services for financial support. Special acknowledgements are made to miss Linda Stannard for the electron microscopy of the viruses and to Mr. D. Engelbrecht, Dr. J. Joubert and Mr. D de Wit for providing the infected plant tissue.

REFERENCES

1. R. E. Hampton and R. W. Fulton, Virology, *13*, 44–52 (1961).
2. A. Polson and L. M. Stannard, Virology, *40*, 781–791 (1970).
3. A. Polson and K. J. Kaufmann, in "Protides of the Biological Fluids", Vol. 18, H. Peeters, eds. Pergamon Press, Oxford, 1970, pp 445–454.
4. A. Polson, in "Methods in Virology", Vol. V, K. Maramorosch and H. Koprowski, eds. Academic Press, New York and London, 1971, pp 33–77.
5. A. Polson and Anne M. Linder, Biochim. et Biophys. Acta, *11*, 119-208 (1953).
6. W. J. K. van der Walt and D. J. Engelbrecht, Phytophylactica, *6* (in press).
7. D. J. Engelbrecht and W. J. K. van der Walt, Phytophylactica, *6* (in press).

20. Electro-Extraction of Viruses from Infected Plant Tissue (Applied to Turnip Yellow Mosaic, Tobacco Mosaic and Maize Streak Viruses)*

ABSTRACT

An apparatus is described which was used for rapid extraction of viruses from frozen and thawed infected plant tissues. The novel principle is the establishment of a potential gradient of 15 to 20 volts/cm at approximately 90° across the leaves of the infected plant while the leaves are surrounded by buffer of low molarity and of the appropriate hydrogen ion concentration. To keep the leaves in the correct orientation they were placed as single layers between coarse rigid plastic gauze. The method, termed electro-extraction, was used as the initial step in the purification of turnip yellow mosaic, tobacco mosaic and maize streak viruses. An electron micrograph of the purified maize streak virus is presented.

INTRODUCTION

In a previous communication to this journal[1] it was shown that viruses could be extracted from infected leaves by subjecting the tissue to a potential gradient of approximately 3.5 volts/cm for 24 hours while the leaves were held with their midribs parallel to the wall of a containing tube in the

*From: A. Polson and K.T. van der Merwe, Prep. Biochem., *11*: 321-338 (1981).

presence of a suitable buffer. The virus particles which migrated out of the tissue were collected in a specially designed cup at the lower end of the tube. The cup was bounded by a semi-permeable membrane to retain the virions. There were disadvantages inherent to the tube method. These were, the limited mass of plant tissue which could be processed in a single tube and the extended time required before the bulk of the virions could be extracted. To improve the method and to make electro-extraction more attractive as a means of obtaining the initial crude extract without disintegration of the plant tissue, a method was devised in which a relatively high electrical potential gradient was established at right angles to the surfaces of the infected leaves while the leaves were held between layers of thin plastic gauze. The virions therefore, only had to migrate a distance equivalent to the thickness of a leaf to escape into the surrounding buffer, leaving the bulk of the virus depleted plant tissue between the gauze sheets.

MATERIAL AND METHODS

The Apparatus

The new apparatus used for electro-extraction is shown diagrammatically in its assembled form in Fig. 1. The two sections "e" are electrode chambers supplied with platinum (Pt) wire electrodes. The electrodes were connected to a source of direct current of which the potential difference between the electrodes could be varied from 0 to 230 volts. The connections to the power supply are indicated by the + and - signs at the top of the assembled apparatus. The "b's" are the sections through which chilled buffer could be circulated if necessary; "x" is the extraction chamber. The different sections are kept apart with wide cellophane membranes. Depending upon the mass of plant tissue to be extracted several extractions compartments, each containing 200 to 300 g leaves may be inserted in place of x, omitting the semipermeable membranes between the sections containing the tissue. Protruding O rings in the flat surfaces of the different sections ensure watertight sealing.

The different compartments are held together with 8 horizontal rods of "tufnol" and 8 steel screws. The chambers were cut from a slab of "teflon" 20 mm thick. The dimensions of the assembled apparatus with one extraction chamber, for the processing of 200-300 g leaves or tissue are 360x160x100 mm.

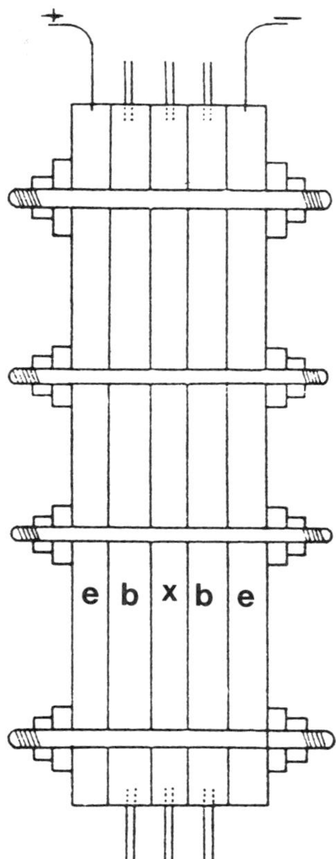

Figure 1. The assembled electro-extraction apparatus. The e's are the electrode, b's the buffer coiling and x the extraction chambers respectively. Cellophane membranes and O rings separate the chambers from one another. The bolts and nuts holding the apparatus together are shown.

The interior of one of the two electrode chambers is shown in Fig. 2 indicating the manner in which the Pt wire (vertical lines) is spaced to ensure approximately uniform distribution of the current through the apparatus. The three horizontal bars in the electrode vessel are teflon rods which keep the Pt windings apart and vertical. The hole, O, near the top corner is used for introducing buffer into the electrode vessel and also to act as outlet for gasses which evolve during electrolysis. The sealing O ring in

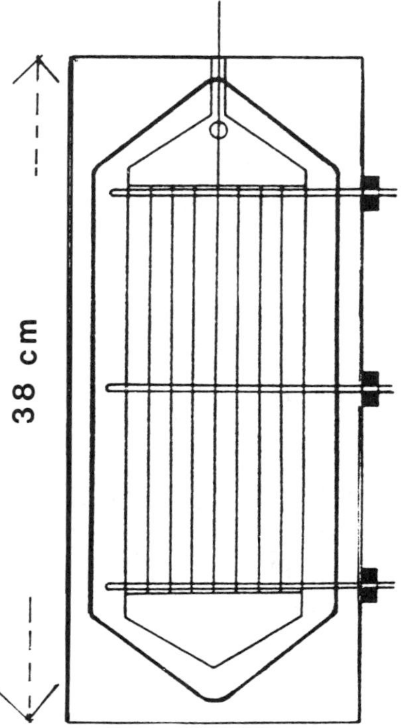

Figure 2. An electrode chamber in a hollowed out section. The hole in the upper corner is for filling the chamber with buffer and is the exit for gasses generated on the electrode.

the frame of the electrode vessel is indicated by the heavy dark line. A sketch of the extraction compartment showing the position of a layer of plant leaves on thin plastic gauze may be seen in Fig. 3. Teflon tubes in the lower extremity and top corner are for introducing dilute buffer and for removal of the extract; also for monitoring the temperature during electro-extraction. The sealing O ring is indicated by the heavy dark line on the frame of the chamber. The buffer cooling chambers ("b") Fig. 1 are identical to Fig. 3 but without the gauze and leaves.

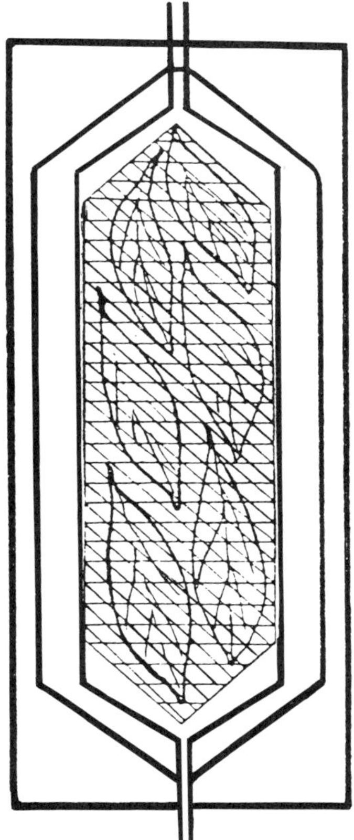

Figure 3. An extraction chamber with a pile of single layered leaves on plastic gauze. The buffer cooling sections are similar to the extraction chamber but without the pack of infected tissue and gauze.

Buffer

It is necessary to use buffers of low conductivity to avoid excessive heat generation during extraction. It is also important to have anti-oxidants such as ascorbic acid and sodium metabisulphite, 0.002 M each, in the buffer; dithiothreitol may also be added. The presence of sodium citrate is advantageous provided the conductivity of the buffer is not increased excessively by the additional salt. Tween 80 at a concentration of 0.1% and/or bovine serum albumin at the same concentration in the buffer was found to be effective virus stabilizing agents.

The extract, which usually was of the order of 250 to 300 ml from one extraction chamber, was freed of finely suspended matter including bacteria by centrifugation at 12 000 g for 15 min. The virions were finally concentrated into pellets by centrifugation at 108 000 g for 90 to 120 min depending upon the estimated sedimentation coefficient of the virus. Centrifugation in the thin layer rotor[2] was also used for obtaining concentrated suspensions of the virions as an alternative to excessive pelleting. Pelleting often causes disruption of virus particles.

Operational

To perform an extraction of virus from infected leaves a pack of thin rigid gauze cut to fit the interior of an extraction chamber and separated from one another by thin layers of infected leaves, was tied together with cotton thread and frozen at -20°C. The different compartments were assembled with cellophane membranes separating the electrode chambers from the buffer cooling compartments. The buffer cooling compartments in turn were separated from the extraction sections respectively. The stack of plastic gauze and frozen leaves was inserted into the extraction chamber. The compartments were clamped together with the tufnol bars and bolts and screws and all five sections filled with very dilute buffer (0.002 M) cooled to 2 to 4°C. A potential of 150 to 230 volts was established across the electrodes which represented a potential gradient of 15 to 23 volts/cm. This resulted in a flow of 0.8 to 1.5 Amp through the apparatus. The current was reversed after 10 min for a further 10 min. Following two additional reversals for 10 seconds at a time the buffer surrounding the plant tissue was drained through the tube intended for this purpose. It is important to avoid overheating as it may happen that during passage of the current the temperature of the extraction chamber's contents could rise to above the thermal inactivation level of the virus. When the temperature of the contents of the extraction chamber had reached 25 to 30°C the extraction was terminated.

Virus Infected Plant Tissue

Chinese cabbage leaves infected with turnip yellow mosaic virus, (TYMV) was obtained from the Institute of Plant Protection, Department of

Agriculture and Fisheries Technical Services, Stellenbosch, tobacco plants infected with the common strain of tobacco mosaic virus (TMV) from the Department of Microbiology, University of Stellenbosch and barley seedlings infected with maize streak virus (MSV) from the Department of Microbiology, University of Cape Town.

RESULTS

Turnip Yellow Mosaic Virus

Electro-extraction was performed on 200 g frozen and thawed TYMV infected Chinese cabbage leaves held in 0.002 M phosphate buffer, pH 7.2 to which the anti-oxidants and protein stabilizing agents were added. The extract, measuring 250 ml, was clarified at 12 000 g for 10 min. The clear supernatant fluid (SNF) was spun at 108 000 g for 120 min in the No 40 Spinco rotor. The pellets were finally dispersed in 1.0 ml 0.02 M phosphate buffer pH 7.2 and examined in the Model E at 25 000 rpm operating at 20°C. Photographic exposures were made at 8 min intervals. The sedimentation diagrams indicated the "top" and "bottom" components originally observed by Markham and Smith[3]. The sedimentation coefficients calculated, 46 and 104 S, respectively were in fair agreement with Markham and Smith's values of 49 and 106 S.

Ouchterlony gel double diffusion tests were made on serial two-fold dilutions of the virus concentrate against rabbit anti TYMV serum. Single precipitin lines were obtained with all the dilutions and the titre observed was in excess of 256.

To indicate that the virions were electro-extracted from the frozen and thawed leaves and that they did not escape from the tissue by diffusion, an extraction chamber was packed with 200 g infected, frozen and thawed leaves and filled with 0.002 M phosphate buffer of pH 7.2. After 20 min the buffer was drained off and the virus which did diffuse out received the same treatment as the electro-extracted material. The final pellet, which was barely visible in the centrifuge tube, was dissolved in 1.5 ml phosphate buffer 0.02 M ph 7.2 and titrated by Ouchterlony double gel diffusion; a titre of approximately 4 was obtained. This experiment provided conclusive evidence that electro-extraction was effective in moving the virions from the tissue into the buffer.

After an electro-extraction of infected Chinese cabbage the tissue in the extraction chamber was rinsed once with the buffer used in the electro-extraction, refilled with chilled buffer and the electro-extraction repeated.

The extract was treated in the same manner as the initial. The final pellet obtained was small and it was estimated that 80 to 90% of the electro-extractable virions was freed from the tissue during the initial extraction.

Tobacco Mosaic Virus

Three hundred grams of frozen and thawed tobacco leaves infected with the common strain of TMV were put through the electro-extraction, concentration and purification as with the purification of TYMV. A small amount of the final pellet obtained by centrifuging the virus suspension at 108 000 g for 90 minutes was mixed with the negative contrast stain potassium phosphotungstate, pH 7.2, and examined in the electron microscope. The electron micrograph obtained indicated that the virions were mainly 300 nm long with few short particles and some virus rods which joined end to end. In general the TMV particles were of the same quality as those obtained by the tube method[1] of electro-extraction.

Maize Streak Virus

Approximately 300 g of barley leaves infected with MSV were frozen at -20°C. Prior to freezing the pack of plastic gauze and leaves was moistened with the "electro-extraction" buffer; acetate of molarity 0.001, pH 4.9 containing 0.001 M sodium meta bisulphite, 0.005 M ascorbic acid and 0.001 M sodium citrate. The moistened pack was held in a plastic bag during freezing. The MSV virions and soluble proteins were extracted from the infected barley leaves as was done with TYMV. The duration of extraction was 25 min, the voltage gradient 15 volts/cm and 1.5 Amp. The temperature of the material in the extraction chamber rose from 4°C to 25°C during extraction. The extract had a total volume of 150 ml. This fluid was freed of coarse matter and bacteria by centrifugation at 12 000 g for 20 min. The clear S.N.F. was concentrated by ultracentrifugation in the thin layer rotor[2] at 25 600 g for 2 hours. The concentrate, measuring 1.5 ml, was diluted to 11 ml in 0.02 M acetate buffer pH 4.9 containing 0.005 sodium meta bisulphite, 0.005 M sodium citrate 0.01 M ascorbic acid and 0.1% Tween 80. After clarification the virus suspension was spun at 120 000 g for 2 hours in the No. 40 Spinco preparative rotor. A clear pellet measuring 6 mm in diameter was obtained. In Fig. 4 an electron micrograph is given of the material after positive contrast staining with uranyl acetate at pH 5.0. The preparation appeared to be free of extraneous matter. The electron micrograph obtained of the virus was typical of the

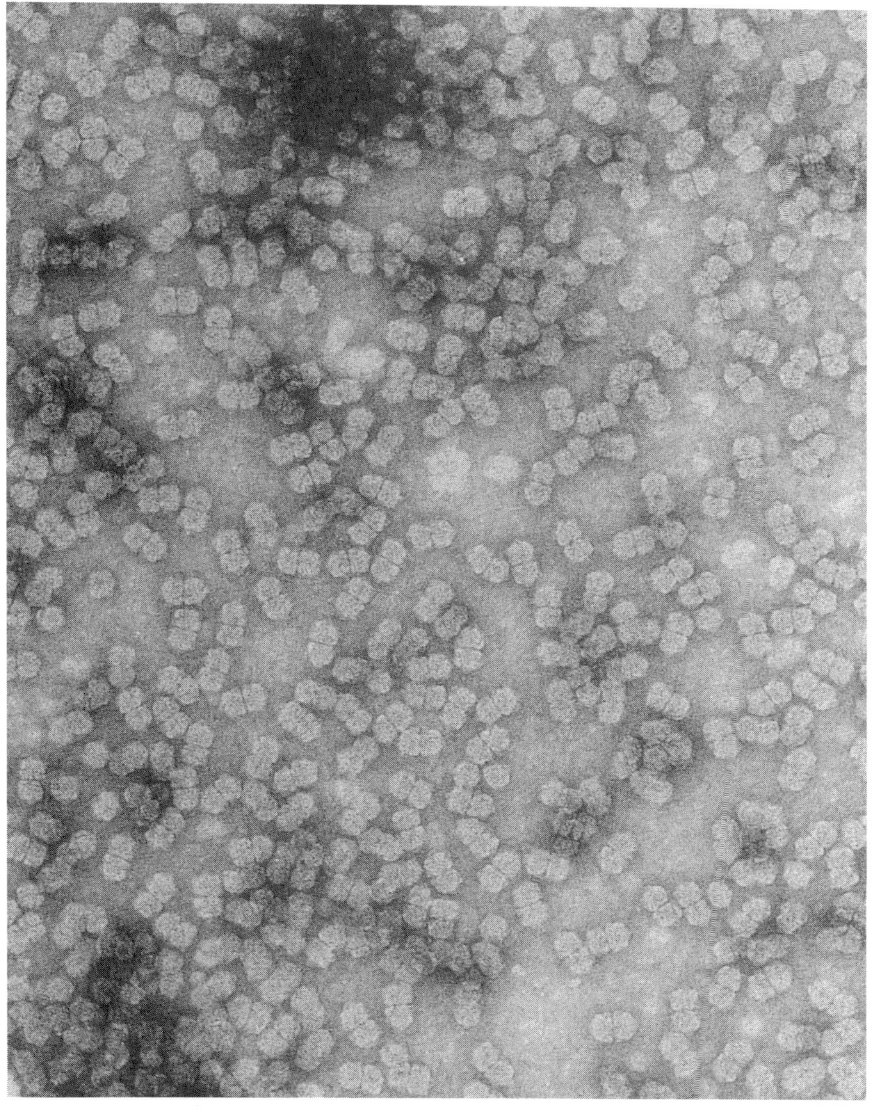

Figure 4. Electron micrograph of maize streak virus electro-extracted prior to purification. The virus was positively contrast stained with uranyl acetate at pH 5.0. Magnification 140 000 x.

Gemini class of viruses and identical to the electron micrograph of MSV presented by Bock et al.[4] The doublets measured 30 by 20 nm.

The yield of MSV virions from 300 g infected barley seedlings in which the electro-extraction procedure was used as the initial step was compared with that obtained when the same mass of unfrozen infected seedlings were crushed in a meat mincer as the initial step. This disintegration of the tissue was done in the presence of 100 ml acetate buffer, 0.02 M pH 4.9 to which the necessary anti-oxidants and Tween 80 were added. The pulp was centrifuged at 25 600 g for 30 min in the bowl of the reorienting gradient rotor fitted with the virus extraction insert[5]. The clear fluid was processed in the same manner as the electro-extracted material. The final pellet obtained after ultracentrifugation at 126 000 g for 2 hours had the same diameter in a tube of the No. 40 Spinco preparative rotor as the virions obtained by electro-extraction as the initial step and was equally free of extraneous matter as revealed by electron microscopy.

DISCUSSION

An improved electro-extraction apparatus capable of bulk separation of virions and soluble substances from infected plant tissue is described. The novel principle involved is the forced migration of these substances into dilute buffer surrounding the tissue by a potential gradient of 15 to 20 volts/cm at 90°C to the surface of frozen and thawed infected leaves. The leaves were held in position as single layers between rigid plastic gauze sheets. It was estimated that more than 80% of the electro-extractable virions could be separated from the plant tissue in 20 min. while the temperature of the mixture being extracted, was well below the thermal inactivation level of most plant viral pathogens.

The technique was used as the initial step in obtaining pure turnip yellow mosaic virus (TYMV) and tobacco mosaic virus (TMV). Phosphate buffer of pH 7.2 and molarity 0.002 to which the necessary anti-oxidants and sodium citrate were added was used as the "extraction" buffer for (TYMV) and (TMV).

As TYMV and TMV are easily purified by the conventional methods, the isolation of pure maize streak virus (MSV) was attempted as the yield of pure MSV by the common procedures of purification could be zero to very little (personal experience - A.P.). When electro-extraction of barley seedlings infected with MSV into acetate buffer of low molarity and pH 4.9 to which the necessary anti-oxidants and Tween 80 were added, no problems were encountered and what appeared to be a good yield of pure MSV resulted.

The extraction and purification of MSV was also accomplished when disintegrated infected barley seedlings were extracted and clarified in the bowl of a re-orienting gradient rotor fitted with a special virus extraction and clarification insert. The final yield of MSV virions appeared to be similar to that obtained following initial electro-extraction.

The successful purification of the virions of MSV could be due to the use of acidic buffer containing the necessary anti-oxidants and, that barley seedlings were used as source of the virus in place of maize (private communication Dr. M.B. von Wechmar, Microbiology, University Cape Town).

ACKNOWLEDGMENTS

We acknowledge gratefully the receipt of tobacco leaves infected with TMV from the Department of Microbiology, University of Stellenbosch; Chinese cabbage infected with TYMV from Plant Protection, Department of Agriculture and Fisheries, Stellenbosch and barley seedlings, infected with MSV, from the Department of Microbiology, University of Cape Town.

The electron micrographs were made by Mr. G. Kasdorf of Plant Protection to whom we are most grateful.

REFERENCES

1. A. Polson, Prep. Biochem. *7*, 207-215 (1977).
2. A. Polson and A. Kiefer, Prep. Biochem. *5*(3), 199-218 (1975).
3. R. Markham and K.M. Smith, Parasitology, *39*, 330-342 (1949).
4. K.R. Bock, J. Guthrie and R.D. Woods, Ann. Appl. Biol., *77*(3), 289-296 (1974).
5. A. Polson, Prep. Biochem., *9*, 427-439 (1979).

21. Purification of Potato Leaf Roll Virus*

A. Polson
Department of Biochemistry, University of Stellenbosch
and
Infertility and Reproductive Biology Unit
Department of Obstetrics and Gynecology
University of Stellenbosch, Tygerberg Hospital
Tygerberg, Republic of South Africa

ABSTRACT

Potato leaf roll virus was purified by a combination of Takanami and Kubo's method and the use of the extraction and thin layer rotors and finally by centrifugation in tubes filled with beads.

INTRODUCTION

Potato leaf roll virus (PLRV) is equally important to the potato industry as the other viruses, potato virus Y° (PVY°) and potato virus X (PVX). Because of the spherical shape of the potato leaf roll (PLRV) virus, it is less affected by such practices which tend to damage the flexous viruses PVY° and PVX. The purification of PLRV was partially based on the method of Takanami and Kubo[1]; the main difference being the use of the virus extraction rotor and centrifugation in centrifuge tubes filled with polyacetate beads.

*Hitherto unpublished.

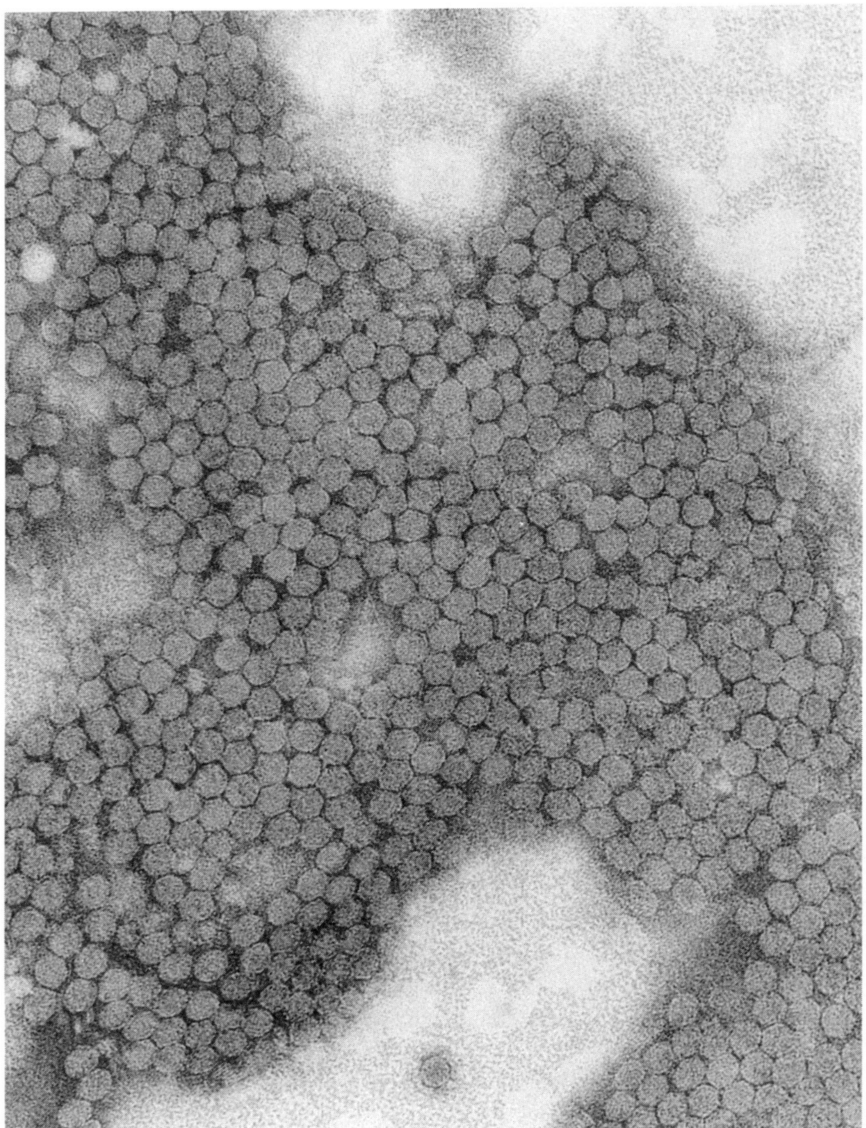

Figure 1a. Electron micrograph of potato leaf roll virus; negatively stained with phosphotungstate magnification 210 000.

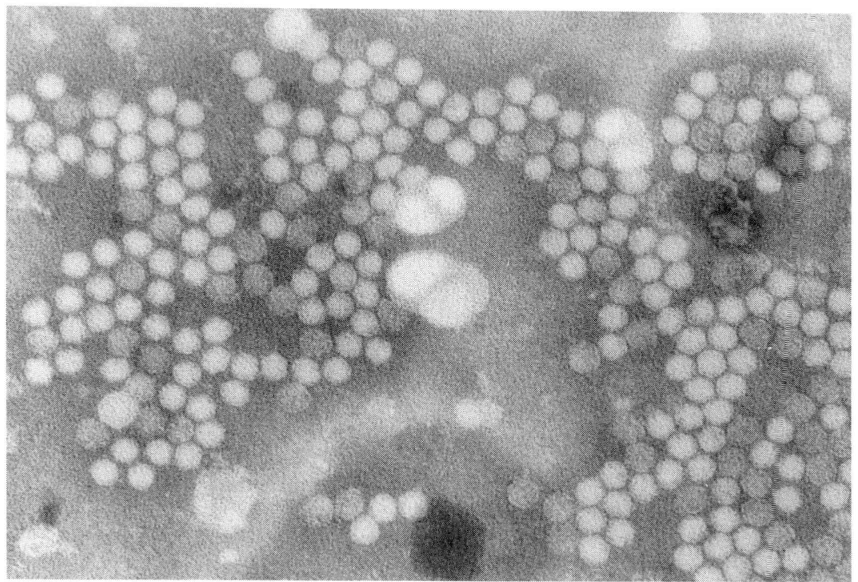

Figure 1b. Electron micrograph of a second isolate of potato leaf roll.

MATERIAL AND METHODS

Infected potato leaves were kindly supplied by Mr. Prins of Rhoodeplaat, Pretoria. On the first consignment of 550g frozen leaves, 137 g were extracted following Takanami and Kubo's[1] procedure. The leaves were disintegrated in a meat grinder in the presence of 300 ml medium containing 100 mM citric acid, 100 mM Na_2HPO_4, pH 5.9 and 0.5% 2 mercapto ethanol. In the disintegrated leaves (pulp) were dissolved 2 g cellulase and 2 g pectonase supplied by Seravac, Cape Town. The mixture was incubated in a water bath at 25°C for 2 hours. The pH was increased to 7.0 by the addition of Na_2HPO_4. At this stage the digested plant material was a slurry. The slurry was centrifuged in the virus extraction rotor. The extract was shaken vigorously with two volumes of butanol-chloroform (1:1) mixture and centrifuged in the J rotor at 5000 g for 10 min.

The watery phase was separated and the virus precipitated with 10% polyethylene glycol (PEG). The virus precipitate was centrifuged off, the supernatant discarded and the precipitate dissolved in 10 ml 100 mM phosphate buffer pH 7.4. After clarification at 2000 g for 10 min the virus was ultra centrifuged off in tubes of the Beckman fixed angle rotor in tubes filled with beads of 3 mm diameter. The rotor was spun at 105 000 g for 20 min.

The virus formed a clear pellet of three mm diameter. Electron microscopy indicated that the pellet was of virus only as no extraneous matter was seen on the micrograph.

Having established that the modifications to Takanami and Kubo's method in the form of the ultracentrifuge rotors may be successfully applied to the remaining infected potato plants 413g (=550 - 137g) the plants were disintegrated in a meat grinder in the presence of 900 ml citric acid - sodium phosphate buffer, pH 5.9 and 0.5% mercapto-ethanol mixture. Four grams of both cellulase and pectonase were added to the plant tissue and incubated for 2 hours in a waterbath at 25°C. The pH of the digested slurry was increased to 7.0 with Na_2HPO_4. The slurry was ultra centrifuged in the virus extraction rotor with the necessary filters fitted. The charged rotor was spun at 15 000 g for 30 min. When the rotor had stopped the SNF was syphoned off. Because of the large volume of slurry it was necessary to extract and clarify the slurry as three separate batches at 15 000g for 30 min at a time. The combined SNFs were pooled and shaken vigorously with an equal volume of chloroform - butanol 1:1 mixture and centrifuged in a Sorvall centrifuge at 5000 g for 10 min. The watery phase was separated and the virus displaced (precipitated) with 10% PEG. The precipitate was centrifuged off and dissolved in 50 ml 100 mM phosphate buffer pH 7.4. After clarification at 2000g for 10 min the virus suspension was divided into 10 portions of 5 ml each and spun at 105 000g for 20 min in tubes filled with polyacetate beads of 3 mm diameter. After decanting the beads and SNF, clear pellets 203 mm diameter were obtained in each of the 10 tubes. A small amount of a pellet was suspended in phosphotungstate and subjected to electron microscopy.

The electron micrograph of the virus is depicted in Fig. 1 showing the virus as a two dimensional lattice and no extraneous matter.

The PLRV was also purified strictly according to the procedure described by Takanami and Kubo starting with 413 g infected plant tissue. The yield was 0.43 of that obtained in the present work judging from absorbance at 260 nm. The reason for the greater yield may very likely be ascribed to the more efficient initial extraction for which the extraction rotor was used. The yield of virus in the present work could have been greater had the remaining tissue on the set of filters been suspended as a thick slurry in a small volume of buffer and reextracted in the rotor.

In conclusion it may be stated that digesting the disintegrated plant tissue with the mixture of cellulase and pectonase as recommended by Takanami and Kubo[1] is essential as without treatment with the enzymes the yield of virus is a small fraction of that when the tissue was given the enzyme digestion treatment.

ACKNOWLEDGMENTS

The author is grateful to the South African Potato Board for financial aid and to the University of Stellenbosch for providing the research facilities. Grateful acknowledgments are also extended to Dr. Linda Stannard of the MRC Virus Research Unit, Medical School, Cape Town and Mr. G. Kassdorf of Plant Protection, Department Agriculture, Stellenbosch for the electron micrographs.

REFERENCE

1. Takanami, Y. and Kubo, S. J. Gen. Virology, *44*, 153-159 (1979).

22. Purification of PVY° and Its Soluble Antigen by Physico Chemical Means*

A. Polson
Department of Biochemistry, University of Stellenbosch
and
Infertility and Reproductive Biology Unit
Department of Obstetrics and Gynecology
University of Stellenbosch, Tygerberg Hospital
Tygerberg, Republic of South Africa

ABSTRACT

Potato virus Y° was purified by centrifugation of infected and minced plant tissue in the virus extraction rotor. As the initial seeding material was heavily contaminated with tobacco mosaic virus (TMV) this virus was isolated and antibody was elicited in chickens. The chicken antibody (IgY) against TMV was used for removing this extraneous virus from the original PVY° seeding material prior to propagating PVY° in tobacco plants, CV Glutinosa.

*Hitherto unpublished.

INTRODUCTION

The purification of filamentous viruses is described in the communication by Polson and Kiefer[1] . In that paper emphasis was laid on the advantages of using the thin layer rotor in the purification of filamentous or flexous viruses. Since the appearance of that paper the virus extraction rotor was introduced for separating viruses from infected plant tissue free of plant debris. Also, the use of centrifuge tube inserts was introduced for reducing the time required for centrifuging a virus from suspension. As the application of the two methods had the potential of extracting more virus from a given mass of infected plant tissue and of concentrating the extracted virus in a greatly reduced time of ultracentrifugation respectively than by the conventional method it was decided to describe the purification of PVY°. The manner of purification of this virus may then serve as a model for purification of other flexous viruses of delicate structure.

Potato virus Y° (PVY°) is of considerable economic importance to the potato industry and the isolation and purification of the virus with the view of eliciting antibody in suitable animals for identification of the virus in potato tubers or leaves by immunological means is the only reliable means of identification of the virus. The immunological technique most commonly used is termed enzyme linked immuno sorbent assay (ELISA). Precipitin reactions and immuno-electron microscopy also supply informative results. In this communication the results obtained mostly with unacquainted techniques are presented as applied to the purification of PVY° and its subunits or soluble antigen. The methods used will only be referred to as they are described in fair detail in this issue of the journal.

MATERIAL AND METHODS

The "seeding" material used for the production of the virus was heavily contaminated with an unidentified strain of tobacco mosaic virus (TMV) and it was necessary to isolate this virus and elicit an antibody against the contaminating agent and to use the antibody for removing the TMV from the PVY°. The procedure followed was to inoculate the contaminated seeding material in tobacco plants CV Samsun as this tobacco cultivar favors the growth of TMV. After three passages of the virus in the cultivar the TMV was free of PVY° as evidenced by electron microscopy using the dipping technique.

Purification of the TMV and Production of Chicken IgY Type Antibody Against the TMV Strain

The tobacco plants (CV Samsun) infected with the extraneous TMV (3rd passage) were minced in a meat grinder in the presence of citric acid - Na_2HPO_4 buffer pH 7.5, EDTA, mercapto-ethanol and DECA. The pulp was centrifuged in the virus extraction rotor at 8000 g for 10 min and the clarified extract shaken with an equal volume of chloroform. The emulsion was centrifuged once more at 1000 g for 10 min and the watery phase once more treated with chloroform and centrifuged at 1000 g for 10 min. The watery phase was spun in the thin layer rotor at 10 000 g for 90 min. The concentrate behind the serrated cord was diluted with 120 ml phosphate buffer 100 nm pH 7.6 and once more centrifuged in the thin layer rotor. The concentrate contained in 2 ml had a blue grey appearance typical of concentrated TMV. The virus concentrate was free of PVY° particles as evidenced by electron microscopy and was used for eliciting IgY type antibodies in chickens (Rhode Island Red X Leghorn). The birds were injected at weekly intervals with 1 mg TMV emulsified in Freund's incomplete adjuvant. Eggs laid by the chickens were collected after the fifth injection and the IgY antibody separated from the yolk using the polyethylene glycol procedure.

Separation of the TMV from the PVY° Seeding Material

Having elicited IgY anti TMV antibody the seeding material was suspended in 10 ml 100 mM phosphate buffer pH 7.2 and increasing amounts of IgY anti TMV added. After each addition the mixture was incubated at room temperature for 30 min. The TMV-IgY complex was centrifuged at 8 000 g for 10 min. The procedure of addition of IgY, incubation for 30 min at room temperature and centrifugation was repeated until no immune complexes were visible when the mixture was centrifuged. The PVY° being unaffected by the TMV immune IgY was centrifuged into a concentrate in the thin layer rotor. On examination in the electron microscope the concentrate appeared free of TMV particles. An additional precautionary measure for obtaining purified PVY° was taken by culturing the virus in tobacco plants of the Glutinosa cultivar. This cultivar is a poor host for TMV but a good one for PVY° (personal communication Professor M.B. von Wechmar).

Purification Procedure of PVY°

The leaves of tobacco plants (CV Glutinosa) infected with PVY° (300 g) were ground in a meat grinder in the presence of the following stabilizing and reducing agents: citric acid - 200 nM Na_2HPO_4 buffer pH 7.4, 10 mM EDTA, 20 nM 2-mercaptoethanol and 1% Na-diethyl-dithio-carbamate. The volume of the mixture was 100 ml. After mincing the leaves, the pulp was ultracentrifuged in the virus extraction rotor (see above). The rotor was designed for extracting and clarifying the plant extract in a single operation. Details of the rotor are given in a previous paper (this journal). Four hundred ml pulp was spun at 15 000 g for 30 min yielding a clarified extract of 360 ml. Because the PVY° is a long flexous virus it could not be centrifuged into and redispersed from a pellet without extensive breakage of the virions resulting in a loss of infectivity. The virus was centrifuged in the thin layer rotor at 15 000 g for 60 min. Because the volume of infected extract was too large for a single centrifugation it was necessary to limit the batch volume treated in one operation to 120 ml. After the first batch was freed of virus, the SNF was syphoned from the rotor cavity through a valve in the lid using a long truncated syringe needle. A second batch of 120 ml was next introduced into the rotor bowl through the valve in the lid. After centrifugation the SNF was removed and the virus in the remaining extract was concentrated. The SNF was removed as was done before and the lid unscrewed. The SNF on the outer surface of the serrated cord was removed and the cord was removed gently. A "rubber policeman" was used to stir up any virus which was centrifuged onto the wall of the receiving cavity. The fluid was removed with a Pasteur pipette of which the tip had been softened in a flame. This precaution was taken to avoid scratching of the surface of the receiving cavity.

The volume of the fluid recovered from the receiving cavity is usually of the order of 1.5 to 2 ml. The virus from the initial pulp may thus be recovered in a small volume as a suspension. On reextracting the residue on the filters in the rotor, more virus may be separated from the disintegrated plant tissue, but this amounts to a small percentage of the agent in the initial extract.

The method of further purifying the virus was to dilute the virus concentrate to 10 ml of 100 mM phosphate buffer pH 7.6 and to shake the suspension briskly with 10 ml chloroform and to centrifuge the emulsion at 5000g for 10 min. The watery phase which was the top layer in the centrifuge tube and which contained the virus was removed and centrifuged in a tube filled with 3 mm polyacetate beads in the fixed angle rotor at

15000 g for 10 min in a Beckman ultracentrifuge. During concentration of the virus in this manner the beads, because of their lower density than the phosphate buffer, separated slightly from the lower end of the tube leaving a clear pellet on the base of the tube. The beads and SNF could be poured out into a glass beaker. On examining a tiny amount of the pellet mixed with 200 mM phosphotungstate in the electron microscope, no extraneous matter but only the virus particles were seen. The electron micrograph obtained was mainly of virus particles in their undamaged form with some smaller particles which possibly were of soluble virus antigen.

Purification of Zone Electrophoresis in a Sucrose Concentration Gradient Using the H-Tube Electrophoresis Apparatus

The use of the H-tube zone electrophoresis apparatus[2] with a sucrose concentration gradient was investigated as an alternative method for the final purification of the virus, the object being to remove the particles which were thought to be virus subunits or soluble antigen seen on the electron-micrograph. The procedure followed was to infect tobacco plants (CV Glutinosa) with the purified PVY°. When the symptoms of the disease were evident the plants were harvested (750g) crushed in a meat grinder in the presence of the anti oxidants and enzyme inhibitors (see above). The volume of the pulp obtained was of the order of 1 liter. The pulp was divided into three lots of approximately 330 ml each. The different lots were clarified by centrifugation in the virus extraction rotor at 15000 g for 30 min as done before. The three batches of clarified fluid were pooled (measured 900 ml) and mixed with 90 g of pulverized polyethylene glycol. The mixture was stirred until all the polymer had dissolved. The precipitate which formed was centrifuged from suspension at 2500 g for 30 min and the pellets dissolved in 120 ml borate buffer pH 8.6 (see previous paper); any coarse suspended matter was removed by light centrifugation. The clarified suspension of the virus was shaken with an equal volume of chloroform and the emulsion which formed was spun in a Sorvall centrifuge at 1 000g for 30 min. The watery phase on the chloroform-protein emulsion was removed, its volume measured and readjusted to 120 ml with borate buffer. The fluid was centrifuged at 12 000 g for 90 min in the tin layer rotor and the concentrate (1.5 ml) behind the serrated cord collected and subjected to zone electrophoresis (ZE) in the H-tube apparatus. The supporting medium was a linear gradient of sucrose in borate buffer ranging from 5% at the top of the to 50% in the cellophane dialysis cup at the lower end of the column.

Prior to electrophoresis a tiny quantity of phenol red (Ø red) was dissolved in the virus concentrate. This pigment served as indicator for the progress of the ZE. When Ø red reached the cellophane dialysis cup the apparatus was dismounted carefully and the tube containing the gradient column held in a retort stand. A bottle of mercury was tied to the dialysis cup to pull it into a concavity. A drawn capillary was lowered down the column and the contents syphoned out in amounts equivalent to 1 cm column length.

The H-tube ZE apparatus and the method of fractionation of the column may be seen in Fig. 1 and details of the apparatus are given in a previous communication. The UV absorbance of the fractions were determined at 260 and 280 nm as well as their ELISA activities expressed as absorbance at 405 nm. It was thus established that the PVY° had an RØ of 0.9. The ELISA activity was absent in the region of the gradient column through which the PVY° migrated. The electron micrograph of the purified virus is presented in Fig. 2. In Fig. 3 an electron micrograph of PVY° "decorated" with its IgY antibody is given.

The PVY° Soluble Antigen

During the initial stages of research on PVY° it was observed that considerable ELISA activity remained in the supernatant fluid (SNF) after the virus had been ultracentrifuged from an extract of infected tobacco plants (CV Glutinosa). It became clear that the activity remaining in the SNF was that of the soluble antigen of the virus. Several ZE experiments were conducted on the soluble antigen in extracts freed of PVY° by ultracentrifugation. The UV absorbance of the fractions at 280 nm as well as the ELISA in terms of absorbance at 405 nm were measured. The RØ of the soluble antigen was found to be 0.6. A ZE experiment was conducted in the H-tube apparatus on the virus free extract using 1% agarose as supporting medium. Following electrophoresis the agarose gel was sliced into sections of 1 cm width which were then frozen, thawed and the containing fluid pressed out. The ELISA activity of the fractions were determined and the data obtained plotted against the fraction number. The diagram again demonstrated that the RØ of the soluble antigen was of the order of 0.6. It became obvious that the soluble antigen could not be freed of extraneous proteins by ZE alone as it was associated with plant proteins. It was therefore decided to separate the soluble antigen by displacement (precipitation) from suspension with polyethylene glycol Mr 6000 (PEG) and to subject the precipitate to ZE. In Fig. 4 a ZE diagram is given of the

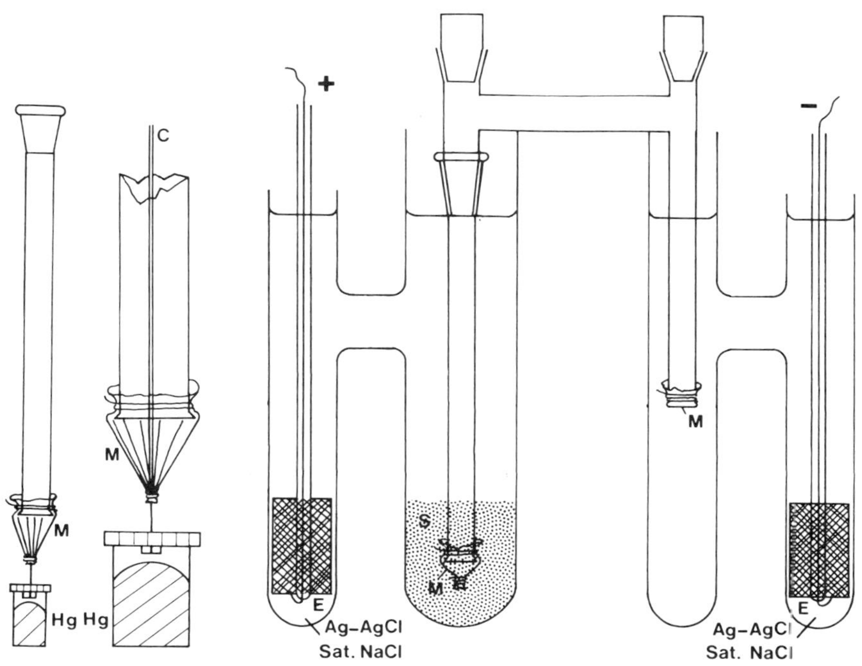

Figure 1. The H tube concentration gradient zone electrophoresis apparatus - details in text.

redissolved precipitate obtained by treatment of the virus free SNF with 10% PEG. The distribution of the ELISA activity is now different, whereas before treatment with PEG the soluble antigen, as evidenced by its ELISA activity, migrated as a single component with an RØ = 0.6, after precipitation with PEG, redissolving in borate buffer and subjected to ZE was no longer homogeneous but the ELISA activity was spread over a large part of the ZE spectrum (Fig. 4a). The main ELISA activity coincided with two protein components of low mobility. The components of lower ELISA activities migrated further.

Apart from the low mobility of the two main ELISA components the "protein" in that region appeared to be aggregated as judged from its opacity.

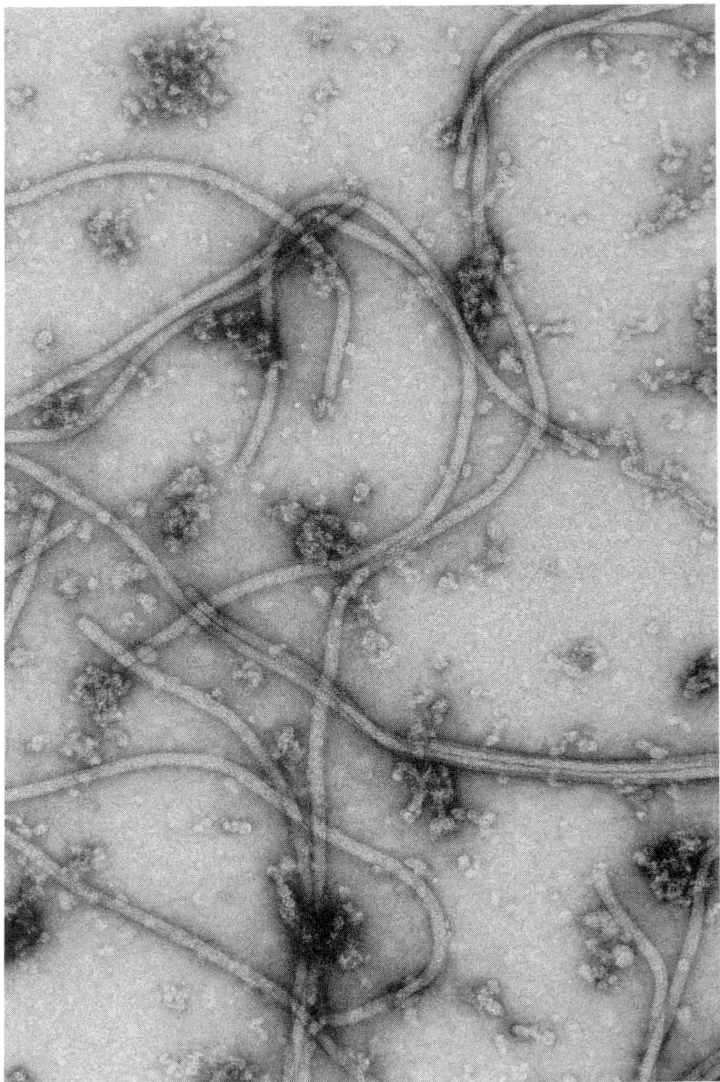

Figure 2. Electron micrograph of PVY° purified by ultracentrifugation, chloroform treatment and zone electrophoresis.

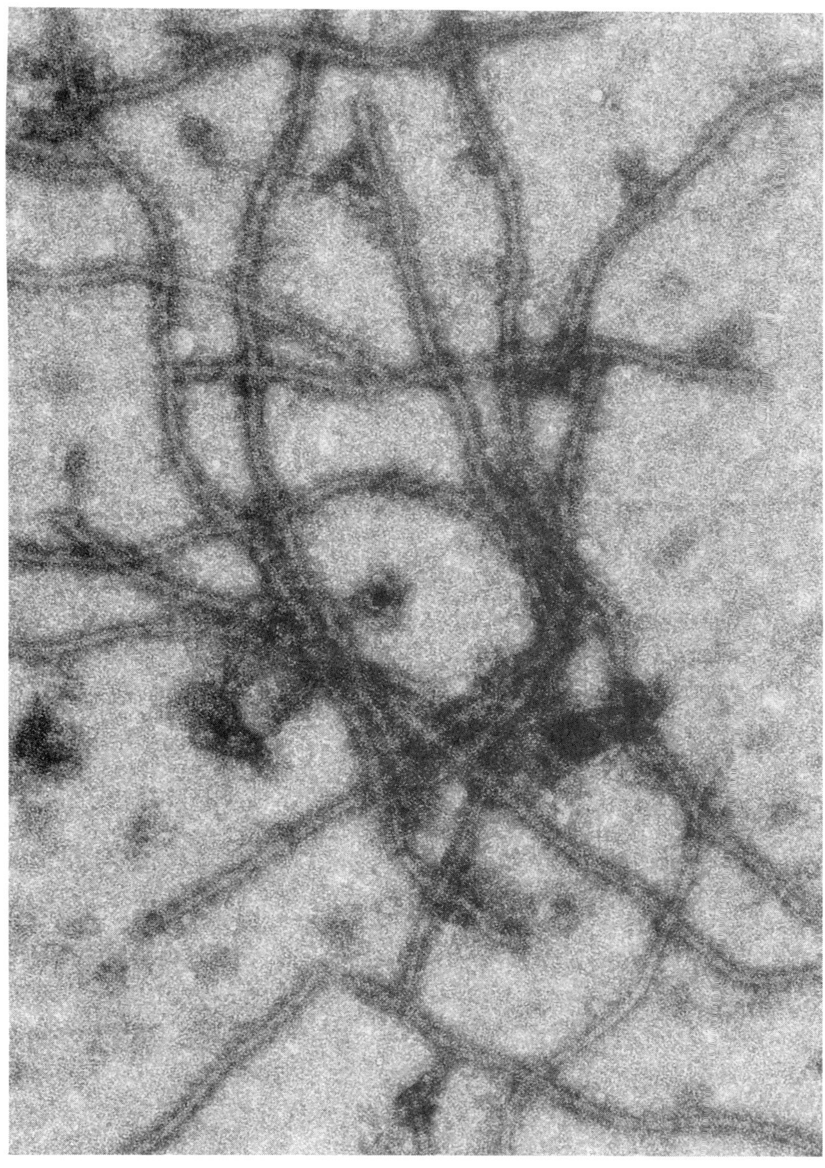

Figure 3. An electron micrograph of PVY° "decorated" with its IgY° (chicken) antibody. Antibody used undiluted.

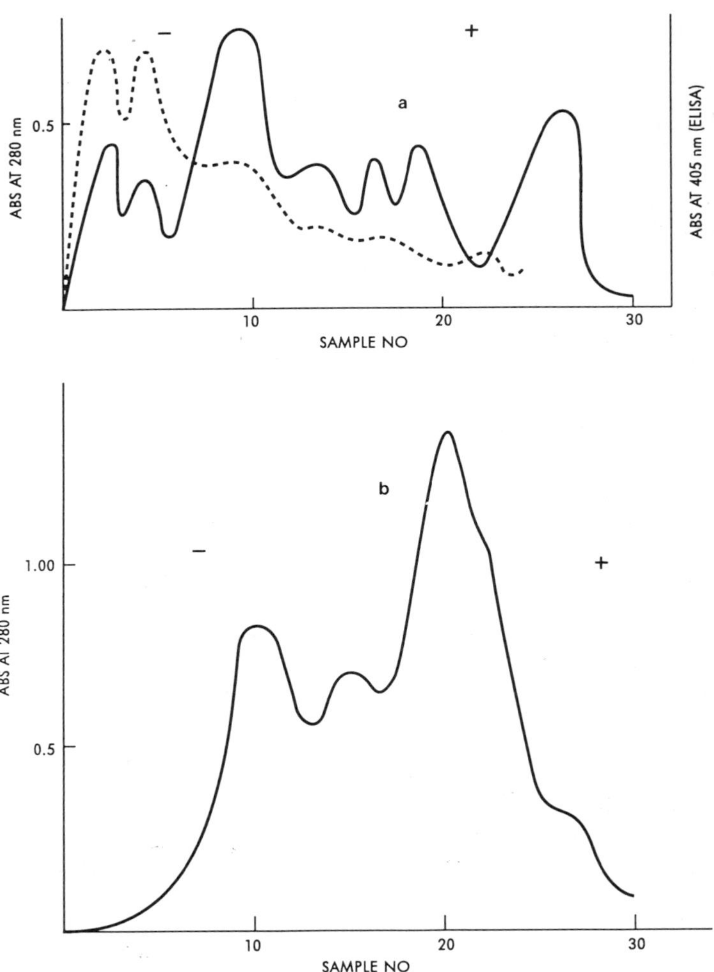

Figure 4. Zone electrophoresis diagram a of the soluble antigen of PVY aggregated by displacement with PEG. The continuous line is the absorbance of the contents of the column at 280 nm. The dashed line is the ELISA activity in terms of absorbance at 405 nm.; b a ZE diagram of uninfected tobacco (CV Glutinosa) extract details in text note the absence of the proteins of the lowest mobilities in Fig. 4a.

The aggregated proteins with the lowest mobilities are the soluble antigens of PVY°. This statement was based on circumstantial evidence. A batch of 750 g normal uninfected tobacco plants (CV Glutinosa) of approximately the same age as the plants infected with PVY° were treated exactly as was done with the infected plants including the precipitation with PEG. The ZE diagram obtained (Fig. 4b) was different in appearance which possibly could be due to the disease but the main difference from the PVY° infected plants is the absence of the two components of lowest mobility with which the main ELISA activity was associated when the soluble antigen was precipitated with PEG.

An attempt at eliciting IgY type antibodies in chickens against the soluble antigen of PVY° was unsuccessful on account of too little antigen available for the full course of immunization.

DISCUSSION AND CONCLUSIONS

Potato virus Y° (PVY°) seeding material was heavily contaminated with an unknown strain of tobacco mosaic virus (TMV) and it was necessary to isolate the TMV in a pure state and elicit antibodies against the purified virus. Because the tobacco cultivar Samsun favors the growth of TMV the PVY° could be suppressed completely after three passages in the Samsun cultivar. Chickens were used for eliciting antibodies against the TMV. To a suspension of the contaminated PVY° seeding material chicken antibody (IgY) to TMV was added incrementally followed by centrifugation at 5000g for 10 min at a time. The procedure was continued until no further TMV-IgY pellets were formed in the centrifuge tubes. The PVY°, which was still in suspension, was now cultivated in tobacco plants (CV Glutinosa) since this cultivar favored the development of PVY°.

The PVY° was purified as shown by electron microscopy and was used as inoculum for the production of IgY type of antibody in chickens. When the purified virus was "decorated" with the IgY antibody the agent reacted with the IgY as shown by electron microscopy. Apart from reacting with the whole virus the IgY also reacted with smaller particles. These were possibly subunits of the virus or what is termed soluble antigen.

An interesting aspect of the soluble antigen is when displaced from suspension with PEG it formed what appeared to be irreversible aggregates. The main aggregates were two entities with low mobilities. They also scattered light strongly which was indicative of low aggregates. The two protein components with which the main ELISA activity was associated are the PVY° soluble antigen because they were absent in the ZE diagram of extracts of tobacco plants (CV Glutinosa) which were not infected with PVY° and which were of the same age. The uninfected plant extract received identical treatment as the infected extracts. Attention is drawn to the fact that when PVY° soluble antigen is displaced with PEG four or possible five ELISA active "aggregates" are formed.

Because the soluble antigen particles aggregate readily when displaced with PEG the inter-particle binding forces would be coulombic in nature and one could speculate that similar forces would be operative intracellularly where PVY° particles are assembled.

ACKNOWLEDGMENTS

The author is grateful to the South African Potato Board for financing this project and to the University of Stellenbosch for the research facilities. Grateful acknowledgments are made to Mr. G. Kasdorf for the electron micrographs.

REFERENCES

1. A. Polson and A. Kiefer, Prep. Bioch., *5*(3), 199-218 (1975).
2. A. Polson, Immunological Investigations, *20*(5 and 6), 451-459 (1991).

Index